This series aims to report new developments in physical research and teaching – quickly, informally, and at a high level. The type of material considered for publication includes:

1. Preliminary drafts of original papers and monographs

2. Lectures on a new field, or presenting a new angle on a classical field

3. collections of seminar papers

4. Reports of meetings

Texts which are out of print but still in demand may also be considered if they fall within these categories.

The timeliness of a manuscript is more important than its form, which may be unfinished or tentative. Thus, in some instances, proofs may be merely outlined and results presented which have been or will later be published elsewhere.

Publication of *Lecture Notes* is intended as a service to the international physical community, in that a commercial publisher, Springer-Verlag, can offer a wider distribution to documents which would otherwise have a restricted readership. Once published and copyrighted, they can be documented in the scientific libraries.

Manuscripts

Manuscripts are reproduced by a photographic process; they must therefore be typed with extreme care. Symbols not on the typewriter should be inserted by hand in indelible black ink. Corrections to the typescript should be made by sticking the amended text over the old one, or by obliterating errors with white correcting fluid. The figures (in the original size) ready for reproduction should be inserted into the text. Should the text, or any part of it, have to be retyped, the author will be reimbursed upon publication of the volume. Authors receive 50 free copies.

The typescript is reduced slightly in size during reproduction, therefore a large size of type should be used; best results will not be obtained unless the text on any one page is kept within the overall limit of 18 x 26.5 cm (7 x 10½ inches). The publishers will be pleased to supply on request special stationery with the typing area outlined.

Manuscripts in English, German or French should be sent to Springer-Verlag, 6900 Heidelberg, Postfach 1780.

Die „*Lecture Notes*" sollen rasch und informell, aber auf hohem Niveau, über neue Entwicklungen in der Physik berichten. Zur Veröffentlichung kommen:

1. Vorläufige Fassungen von Originalarbeiten und Monographien.

2. Spezielle Vorlesungen über ein neues Gebiet oder ein klassisches Gebiet in neuer Betrachtungsweise.

3. Seminarausarbeitungen.

4. Vorträge von Tagungen.

Ferner kommen auch ältere vergriffene spezielle Vorlesungen, Seminare und Berichte in Frage, wenn nach ihnen eine anhaltende Nachfrage besteht.

Die Beiträge dürfen im Interesse einer größeren Aktualität durchaus den Charakter des Unfertigen und Vorläufigen haben. Sie brauchen Beweise unter Umständen nur zu skizzieren und dürfen auch Ergebnisse enthalten, die in ähnlicher Form schon erschienen sind oder später erscheinen sollen.

Die Herausgabe der „*Lecture Notes*" Serie durch den Springer-Verlag stellt eine Dienstleistung an die physikalischen Institute dar, indem der Springer-Verlag für ausreichende Lagerhaltung sorgt und einen großen internationalen Kreis von Interessenten erfassen kann. Durch Anzeigen in Fachzeitschriften, Aufnahme in Kataloge und durch Anmeldung zum Copyright sowie durch die Versendung von Besprechungsexemplaren wird eine lückenlose Dokumentation in den wissenschaftlichen Bibliotheken ermöglicht.

Lecture Notes in Physics

Edited by J. Ehlers, München, K. Hepp, Zürich and
H. A. Weidenmüller, Heidelberg
Managing Editor: W. Beiglböck, Heidelberg

21

Optimization and Stability Problems in Continuum Mechanics

Lectures Presented at the Symposium on Optimization and Stability Problems in Continuum Mechanics
Los Angeles, California, August 24, 1971

Edited by P. K. C. Wang
University of California, Los Angeles, CA/USA

Springer-Verlag
Berlin Heidelberg GmbH 1973

ISBN 978-3-540-06214-1   ISBN 978-3-540-38495-3 (eBook)
DOI 10.1007/978-3-540-38495-3

Originally published by Springer-Verlag Berlin Heidelberg New York in 1973

PREFACE

The five papers in this volume represent expanded versions of the lectures presented at the Symposium on Optimization and Stability Problems in Continuum Mechanics at the University of Southern California, Los Angeles, August 24,1971. The Symposium was held in conjunction with the Western Applied Mechanics Conference sponsored by the Applied Mechanics Division of the American Society of Mechanical Engineers with the cooperation of the University of Southern California.

The objectives of this Symposium were twofold,namely,to introduce recent results in general optimization and stability theories which have potential applications to continuum mechanical systems and to present new results dealing with specific classes of systems. It is felt that there is a wealth of new and interesting optimization and stability problems in continuum mechanics. Hopefully, these lectures will help to stimulate further research in this relatively new area.

The idea for this Symposium was originally conceived by Professor C.S.Hsu of the University of California,Berkeley, who also presided over the Stability Session of this Symposium. Professor G.H.Hegemier of the University of California,San Diego, served as the Co-Chairman of this Symposium.

Los Angeles,California

April,1972

P.K.C.Wang

CONTENTS

PART 1. OPTIMIZATION PROBLEMS

PART 2. STABILITY PROBLEMS

THE METHOD OF DUBOVITSKII-MILYUTIN IN MATHEMATICAL PROGRAMMING*

Hubert Halkin**

Department of Mathematics
University of California at San Diego
La Jolla, California

1. INTRODUCTION

I want to give here a brief description of a very attractive formalism in optimization theory: the method of Dubovitskii and Milyutin [1] and the relate some recent extensions of that method, Halkin [2], with the necessary condition of Fritz John[3].

The first step in the method of Dubovitskii-Milyutin is to notice that, in any optimization problem, to say that some solution is optimal is equivalent of saying that a certain family of sets $\{S_i : i \in I\}$ have no points in common i.e. $\bigcap_{i\in I} S_i = \Phi$. Consider for example the optimization problem consisting in minimizing a function f over all point of the plane R^2 where a function g is nonpositive. To say that an element $\hat{x}$ in R^2 with $g(\hat{x}) \leqslant 0$ is an optimal solution for that problem is equivalent of saying that the sets $S_1 = \{x : x \in R^2, f(x) < f(\hat{x})\}$ and $S_2 = \{x : x \in R^2, g(x) \leqslant 0\}$ have no points in common. In general we can do very little with two disjoint sets S_1 and S_2. But if S_1 and S_2 happen to be convex and nonempty then we can separate them by an hyperplane, i.e. we can find a nonzero vector p in R^2 such that

$$\sup_{x \in S_1} p \cdot x \leqslant \inf_{x \in S_2} p \cdot x .$$

Here the scalar product of two vectors p and x is denoted by $p \cdot x$. We shall show below that this separation, when possible, leads to interesting results.

The second step in the method of Dubovitskii-Milyutin is to repace each set S_i by a set Ω_i which is convex and such a good approximation of S_i (in a sense to be precised later) that from the fact that the sets S_i have no point in common, i.e. $\bigcap_{i\in I} S_i = \Phi$, we can prove that the sets Ω_i will have no point in common, i.e. $\bigcap_{i\in I} \Omega_i = \Phi$.

* This research was supported by the Air Force Office of Scientific Research, under Grant No. AFOSR-68-1529C.

**On leave for the academic year 1971-1972 at CORE, University of Louvain, Belgium.

In the simple example given above, if f and g are differentiable at x, we could consider the convex sets

$\Omega_1=\{x : x\in R^2, f(\hat{x})+(x-\hat{x})\cdot \text{grad } f(\hat{x})<f(\hat{x})\}=\{x : x\in R^2, (x-\hat{x})\cdot \text{grad } f(\hat{x})<0\}$ and

$\Omega_2=\{x : x\in R^2, g(\hat{x})+(x-\hat{x})\cdot \text{grad } g(\hat{x})<0\}$.

The third step in the method of Dubovitskii-Milyutin is to go from the fact that the sets Ω_i have no point in common, i.e. $\bigcap_{i\in I} \Omega_i=\Phi$ to the fact that the sets $\{\Omega_i : i\in I\}$ can be separated. We shall define later what we mean by separating more than <u>two</u> convex sets. For the time being let us go back to our simple example where we have only two sets Ω_1 and Ω_2 and let us see how the separation of the two convex sets Ω_1 and Ω_2 is equivalent to the known necessary condition for that problem: if $\hat{x}$ is optimal and grad $g(\hat{x})\neq 0$, then for some $\lambda \leqslant 0$ we have $\lambda g(\hat{x}) = 0$ and $-\text{grad } f(\hat{x}) + \lambda \text{grad } g(\hat{x}) = 0$. The set Ω_1 will always be empty if grad $f(\hat{x}) \neq 0$ and the set Ω_2 will always be empty if grad $g(\hat{x}) \neq 0$. We shall temporarily assume that grad $f(\hat{x})$ and grad $g(\hat{x})$ are different from zero. Since for i=1 and 2 the point $\hat{x}$ belongs to $\overline{\Omega}_i$, (the closure of Ω_i), i.e. since there are points in Ω_i arbitrarily close to $\hat{x}$, then the hyperplane separating Ω_1 and Ω_2 must pass through $\hat{x}$, i.e. there exists a nonzero vector p such that $\sup_{x\in\Omega_1} p\cdot x = p\cdot\hat{x} = \inf_{x\in\Omega_2} p\cdot x$. In other words, for all $x\in\Omega_1$, i.e. for all x such that $(x-\hat{x})\cdot\text{grad } f(\hat{x}) < 0$ we have $p\cdot x \leqslant p\cdot\hat{x}$, i.e. $(x-\hat{x})\cdot p \leqslant 0$. Since $p \neq 0$, this means that for some $a > 0$ we must have grad $f(\hat{x}) = ap$. Moreover, for all $x \in \Omega_2$, i.e. for all x such that $g(\hat{x}) + (x-\hat{x})\cdot\text{grad } g(\hat{x}) < 0$, we have $p\cdot x \geqslant p\cdot\hat{x}$; i.e. $(x-\hat{x})\cdot p \geqslant 0$. Since $p \neq 0$, this means that $g(\hat{x})=0$ and that for some $b>0$, we have $p=-b$ grad $g(\hat{x})$ and $bg(\hat{x})=0$. Combining those two results, we obtain that, under the assumptions grad $f(\hat{x})\neq 0$ and grad $g(\hat{x})\neq 0$, we have $-\text{grad } f(\hat{x}) + \lambda\text{grad } g(\hat{x}) = 0$ and $\lambda g(\hat{x})=0$ for some $\lambda=-ab < 0$. In the case where grad $g(\hat{x}) \neq 0$ and grad $f(\hat{x}) = 0$, we can state trivially that $-\text{grad } f(\hat{x}) + \lambda\text{grad } g(\hat{x}) = 0$ and $\lambda g(\hat{x}) = 0$ by letting $\lambda=0$. We have thus obtained the classical result already mentioned above: if $\hat{x}$ is optimal and grad $g(\hat{x}) \neq 0$, then for some $\lambda\leqslant 0$ we ve $\lambda g(\hat{x}) = 0$ and $-\text{grad } f(\hat{x}) + \lambda\text{grad } g(\hat{x}) = 0$.

Without the assumption grad $g(\hat{x}) \neq 0$, this classical result would be incorrect as one can see in the following simple optimization problem on the real line R^1 : minimize $f(t)\equiv t$ subject to the constraint $g(t)\equiv t^2<0$. The optimal solution is obviously $\hat{t}=0$. But since grad $f(0)=1$ and grad $g(0)=0$, it is impossible to find a real number $\lambda\leqslant 0$ such that $-\text{grad } f(0) + \lambda \text{grad } g(0)=0$, i.e. such that $-1 + \lambda\cdot 0 = 0$. The reader should realize that in this simple pathologival problem the sets Ω_1 and Ω_2 are nevertheless disjoint since Ω_1 is the set $\{t:t < 0\}$ and Ω_2 is the empty set.

The assumption grad $g(\hat{x}) \neq 0$ is the most benign form of a general class of assumptions known as constraint qualifications; I shall come back to that topic in Section 5. If we do not assume that grad $g(\hat{x})\neq 0$, then the necessary condition for that problem takes the form: if $\hat{x}$ is optimal, then for some $(\alpha,\beta)\neq 0$ with α and $\beta\leqslant 0$, we have $\alpha \text{ grad } f(\hat{x}) + \beta \text{grad } g(\hat{x})=0$ and $\beta g(\hat{x})=0$. Indeed if grad $g(\hat{x})\neq 0$, then we let $\alpha=-1$ and $\beta=\lambda$, where λ is the number given earlier; if on the other hand we have grad $g(\hat{x})=0$, then either $g(\hat{x})=0$ or $g(\hat{x})<0$. If $g(\hat{x})=0$, we let $\alpha=0$ and $\beta=-1$. If $g(\hat{x})<0$, then the fact that the sets Ω_1 and Ω_2 are disjoint implies that grad $f(\hat{x})=0$ and in that case we let $\alpha=-1$ and $\beta=0$.

In this paper, I shall always use the n-dimensional Euclidean space R^n as the basic reference space. Although this is sufficient for most applications to mathematical programming, this is not the case in the theory of optimal control in which we must consider infinite-dimensional spaces of trajectories. However, the reader should be aware that everything stated here can be extended to general normed linear spaces, Halkin [2], in which optimal control problems can be treated.

2. SEPARATING SEVERAL CONVEX SETS

If p is a vector in R^n and if α is a real number, then the function f defined over R^n by the relation $f(x)=p\cdot x + \alpha$ is called an affine function on R^n. Colloquially speaking, an affine function is a linear-plus-a-constant function. If $p\neq 0$, then the affine function $p\cdot x + \alpha$ is said to be an affine nonconstant function. A finite family $\{\Omega_i : i\in I\}$ of convex sets will be said to be separated, if there exists a finite family of affine functions $\{\omega_i : i\in I\}$ such that

(i) $\Sigma_{i\in I}\ \omega_i = 0$,

(ii) $\omega_i(x) \geqslant 0$ for all $i\in I$ and all $x\in\Omega_i$,

(iii) ω_i is nonconstant for some $i\in I$.

Let us show that in the case of <u>two</u> sets Ω_1 and Ω_2, this new definition of separation coincides with the classical definition of separation. If Ω_1 and Ω_2 are (classically) separated, then there exists a nonzero vector p such that $\sup_{x\in\Omega_1} p\cdot x \leqslant \inf_{x\in\Omega_2} p\cdot x$. If we let $\omega_1(x)=-p\cdot x + \sup_{x\in\Omega_1} p\cdot x$ and $\omega_2(x)=p\cdot x - \sup_{x\in\Omega_1} p\cdot x$, we thus obtain

(i) $\omega_1 + \omega_2 = 0$,

(ii) $\omega_i(x) \geqslant 0$ for all $i \in \{1,2\}$ and all $x \in \Omega_i$,

(iii) both ω_1 and ω_2 are nonconstant.

In other words, Ω_1 and Ω_2 are separated according to the new definition. Conversely, if Ω_1 and Ω_2 are separated according to the new definition, i.e. if for some $\omega_1(x)= p_1\cdot x + \alpha_1$ and $\omega_2(x) = p_2\cdot x + \alpha_2$, we have

(i) $\omega_1 + \omega_2=0$,

(ii) $\omega_i(x) \geqslant 0$ for all $i\in\{1,2\}$ and all $x\in\Omega_i$,

(iii) at least one of the vectors p_1 or p_2 is different from zero.

Then, by (i), we have $p_2 = -p_1$, $\alpha_2 = -\alpha_1$ and, by (iii), we have $p_2 = -p_1 \neq 0$. If we let $p = p_2 = -p_1$, we obtain $\sup_{x\in\Omega_1} p\cdot x \leqslant \inf_{x\in\Omega_2} p\cdot x$ and the two sets Ω_1 and Ω_2 are separated according to the classical definitions.

We know that two disjoint nonempty convex sets in R^n can be separated, but it is not correct to say that two separated convex sets in R^n are disjoint. For instance the sets $\Omega_1 = \{(x_1,x_2)\colon x_1\leqslant 0\}$ and $\Omega_2 =\{(x_1,x_2)\colon x_1\geqslant 0\}$ in the plane R^2 are not disjoint, since $\Omega_1 \cap \Omega_2 = \{(x_1,x_2)\colon x_1=0\} \neq \Phi$ but they are separated since for $p=(1,0) \neq 0$, we have $\sup_{x\in\Omega_1} x\cdot p = \inf_{x\in\Omega_2} x\cdot p$. If either Ω_1 or Ω_2 is open however, we know that the fact that Ω_1 and Ω_2 are separated will imply that Ω_1 and Ω_2 are disjoint. The same result can be extended to several convex sets in the following manner:

<u>Theorem 2.1</u>. If $\{\Omega_i : i\in I\}$ is a finite family of nonempty convex sets in R^n such that $\bigcap_{i\in I} \Omega_i$ is empty, then the family $\{\Omega_i : i\in I\}$ can be separated. Conversely, if $\{\Omega_i : i\in I\}$ is a finite separated family of convex sets in R^n and if at most one of them fails to be open, then $\bigcap_{i\in I} \Omega_i$ is empty.

The proof of Theorem 2.1 can be found in Halkin [2].

In several applications, we shall assume that $0 \in \overline{\Omega}_i$ for each $i \in I$. In that case, if the family $\{\Omega_i\colon i \in I\}$ of convex sets is separated by the family $\{\omega_i : i\in I\}$ of affine functions, then we shall have $\omega_i(0)\geqslant 0$ for each $i\in I$ and $\Sigma_{i\in I}\,\omega_i(0)=0$ which imply $\omega_i(0)=0$ for each $i\in I$ and hence the functions ω_i are not only affine but linear, i.e. of the form $\omega_i(x)=p_i\cdot x$. We can thus state that a family $\{\Omega_i : i\in I\}$ of convex sets with $0\in\overline{\Omega_i}$ for each $i\in I$ is separated if and only if there exists a finite set of vectors $\{p_i : i\in I\}$ such that

(i) $\Sigma_{i\in I}\, p_i = 0$, (ii) $p_i\cdot x \geqslant 0$ whenever $i\in I$ and $x\in\Omega_i$,

(iii) $p_i \neq 0$ for some $i\in I$.

3. CONVEX APPROXIMATIONS OF SETS

We shall consider three different types of convex approximations of sets: (i) the interior convex approximation (ii) the tangent convex approximation and (iii) the simplicial^k convex approximation where k is a positive integer. In mathematical programming, the two concepts of interior convex approximation (asociated with the objective function and the inequality constraints) and of tangent convex approximation (associated with the equality constraints) are the most useful. The simplicial^k convex approximation is used chiefly in optimal control theory and is associated with operator constraints (i.e. when one requires a trajectory to satisfy some differential equations). However, in the case k=1, i.e. in the case of the simplicial^1 convex approximation, this concept is also used in mathematical programming under the form of the Abadie Sequential Constraint Qualification, Abadie [4].

We should normally give all the definitions under the form: the set Ω is an interior (resp. tangent or simplicial^k convex approximation around a point $\hat{x}$ to a set S, if .... For the sake of simplicity of notation, we shall give all those definitions with respect to the point $\hat{x}=0$. To go back to the general case, we shall use the following convention: the set Ω is an interior (resp. tangent or simplicial^k) convex approximation around a point $\hat{x}$ to the set S, if the set $\Omega-\hat{x}$ is an interior (resp. tangent or simplicial^k convex approximation to the set $S-\hat{x}$). If A is the set in R^n and a is a vector in R^n, then we use the notation A-a to denote the set $\{x-a : x\in A\}$. Let us specify some further notations. If $x\in R^n$, then $|x|$ will be the Euclidean length of x. If $A\in R^n$, then coA will be the convex hull of A. A set $\{x_1,\ldots,x_\ell\}$ in R^n is said to be in general position if the vectors $x_2-x_1, x_3-x_1,\ldots,x_\ell-x_1$ are linearly independent.

Definition 3.1. A subset Ω of R^n is an interior convex approximation to a subset S of R^n if (i) Ω is open, (ii) Ω is convex, (iii) $0\in\bar{\Omega}$ and (iv) for all $\bar{x}\in\Omega$ there exists an $\varepsilon>0$ such that $\eta x\in S$ whenever $|x-\bar{x}|<\varepsilon$ and $\eta\in(0,\varepsilon)$.

Definition 3.2. A subset Ω of R^n is a tangent convex approximation to a subset S of R^n if there exists a neighborhood V of 0 and a continuous real-valued function ϕ defined on V, differentiable at x=0 and such that (i) grad $\phi(0) \neq 0$, (ii) $\phi(0)=0$, (iii) $\Omega=\{x:x\in R^n, x\cdot \text{grad}\,\phi(0)=0\}$ and (iv) $\{x:x\in V,\phi(x)=0\}\subset S$.

Definition 3.3. If k is a positive integer, we shall say that a subset Ω of R^n is a simplicial^k convex approximation to a subset S of R^n, if (i) Ω is convex, (ii) $0\in\bar{\Omega}$ and (iii) for any set $\{x_1,\ldots,x_\ell\}$ with $\ell\leq k$ elements in general position in Ω and for any real number $\varepsilon>0$ there exists a number $\eta\in(0,\varepsilon)$ and a continuous function ζ from $\text{co}\{x_1,\ldots,x_\ell\}$ into R^n such that $|\zeta(x)-x|\leq\varepsilon$ and $\eta\zeta(x)\in S$ whenever $x\in\text{co}\{x_1,\ldots,x_\ell\}$.

Remark: As I mentioned before, the concept of simplicial^1 convex approximation is related to Abadie Sequential Constraint Qualification. Indeed from Definition 3.3, we have: a subset Ω of R^n is a simplicial^1 convex approximation to a subset S of R^n if (i) Ω is convex, (ii) $0\in\bar{\Omega}$ and (iii) for each $\bar{x}\in\Omega$ and each real number $\varepsilon>0$, there exists a number $\eta\in(0,\varepsilon)$ and an element $y\in R^n$ such that $|y-\bar{x}|\leq\varepsilon$ and $\eta y\in S$. The last condition can be rewritten as: for each $\bar{x}\in\Omega$, there exists a sequence of positive real numbers $\eta_1,\eta_2,\ldots$ and a sequence of elements $y_1,y_2,\ldots$ in R^n such that $\lim_{i\to\infty}|y_i-\bar{x}|=0$, $\lim_{i\to\infty}\eta_i=0$ and $\eta_i y_i\in S$ for all $i=1,2,\ldots$.

Examples of Convex Approximations. If ϕ is a real-valued function defined on R^n such that (i) $\phi(0)\leq 0$ and (ii) grad $\phi(0)$ exists and is different from zero, then $\Omega=\{x:x\in R^n, x\cdot\text{grad}\,\phi(0)<0\}$ is an interior convex approximation to each of the sets $S=\{x:x\in R^n, \phi(x)<0\}$ and $\bar{S}=\{x:x\in R^n,\phi(x)\leq 0\}$, (the proof of that fact is not too hard). If moreover, $\phi(0)=0$ and ϕ is continuous in some neighborhood of 0, then $\Omega^*=\{x:x\in R^n, x\cdot\text{grad}\,\phi(0)=0\}$ is a tangent convex approximation to the set $S^*=\{x:x\in R^n,\phi(x)=0\}$, (there is nothing to prove here, just apply the definition). As I mentioned before, simplicial^k convex approximations are used in optimal control theory to handle operator constraint of the type $x\in S$ where S is the set of all trajectories which are solutions of a given family of ordinary differential equations. It is very hard to express operator constraint in terms of inequality and/or equality constraint(s) and even when it is possible the function describing those constraints are not "smooth" enough to apply the concepts of interior convex approximation and/or tangent convex approximation. This is the reason why it is convenient to keep operator constraints under their given forms and to define a special type of convex approxiamtion adapted to those operator constraints. This special type of convex approximation is the simplicial^k convex approximation. In optimal control theory, the simplicial^k convex approximation Ω to the set S will be the set of all solutions of a certain linearization of the given family of ordinary differential equations. For more details, see Halkin [5] and Halkin-Neustadt [6].

4. THE THEOREM OF DUBOVITSKII AND MILYUTIN

<u>Theorem 4.1</u>. Let $S_{-\mu},\ldots,S_{-1},S_0,S_1$ be subsets of R^n such that $\bigcap_{i=-\mu}^{+1} S_i=\phi$. Assume that we have convex sets $\Omega_{-\mu},\ldots,\Omega_1$ such that Ω_i is an interior convex approximation to S_i for each $i=-\mu,\ldots,0$ and such that Ω_1 is a simplicial^1 convex approximation to S_1. Then, the sets $\Omega_{-\mu},\ldots,\Omega_1$ are disjoint and hence separated.

The proof of Theorem 4.1 is particularly simple. If the sets $\Omega_{-\mu},\ldots,\Omega_1$ are not disjoint, then there exists an element $\bar{x}$ which belongs to each of them. Since for each $i\in\{-\mu,\ldots,0\}$ the set Ω_i is an interior convex approximation to the set S_i, we know that there exists an $\varepsilon_i>0$ such that $\eta x\in S_i$ whenever $|x-\bar{x}|<\varepsilon_i$ and $\eta\in(0,\varepsilon_i)$. If we let $\varepsilon=\min\{\varepsilon_{-\mu},\ldots,\varepsilon_0\}$ we see that $\eta x\in S_i$ whenever $i\in\{-\mu,\ldots,0\}$, $|x-\bar{x}|<\varepsilon$ and $\eta\in(0,\varepsilon)$. Since $\bar{x}\in\Omega_1$ and since Ω_1 is a simplicial^1 convex approximation to S_1, we know that there exists a number $\eta\in(0,\varepsilon)$ and an element $y\in R^n$ such that $|y-\bar{x}|<\varepsilon$ and $\eta y\in S_1$. Since $\eta\in(0,\varepsilon)$, the element ηy belongs to every S_i for all $i\in\{-\mu,\ldots,0\}$. This contradiction concludes the proof of Theorem 4.1.

The results of Theorem 4.1 can be applied directly to the following mathematical programming problem: given a set $S_1\subset R^n$ and given functions $\phi_{-\mu},\ldots,\phi_{-1},\phi_0$ defined over R^n, find an element $x\in S_1$ which minimizes $\phi_0(x)$ subject to constraints $\phi_i(x)\leq 0$ for $i=-\mu,\ldots,-1$. We assume that an optimal solution $\hat{x}$ exists for that problem, that Ω_1 is a simplicial^1 convex approximation to S_1 around $\hat{x}$ and that $\phi_{-\mu},\ldots,\phi_0$ are differentiable at $\hat{x}$. We then obtain

<u>Theorem 4.2</u>. Under the preceeding assumptions, there exist numbers $\lambda_{-\mu},\ldots,\lambda_0$, not

all zero, such that

(i) $\lambda_i \leqslant 0$ for each $i \in \{-\mu,\dots,0\}$,

(ii) $\lambda_i \phi_i(\hat{x}) = 0$ for each $i \in \{-\mu,\dots,-1\}$,

(iii) $\Sigma_{i=-\mu}^{0} \lambda_i \text{ grad } \phi_i(\hat{x}) \cdot (x-\hat{x}) \leqslant 0$ for all $x \in \Omega_1$.

Proof of Theorem 4.2: Without loss of generality, we assume that $\hat{x}=0$ and that $\phi_0(0)=0$. For each $i \in \{-\mu,\dots,0\}$,let $S_i=\{x:\phi_i(x)<0\}$ and let $\Omega_i=\{x:\phi_i(0)+\text{grad } \phi_i(0)\cdot x<0\}$. According to Theorem 4.1, the sets $\{\Omega_i : i=-\mu,\dots,1\}$ are disjoint and hence separated. By construction, we have $0 \in \overline{\Omega}_i$ for all $i \in \{-\mu,\dots,+1\}$ and hence there exists a set of vectors $\{p_{-\mu},\dots,p_1\}$, not all zero, such that

(i) $p_i \cdot x \geqslant 0$ whenever $x \in \Omega_i$ and $i \in \{-\mu,\dots,+1\}$,

(ii) $\Sigma_{i=-\mu}^{+1} p_i = 0$.

Since $\Sigma_{i=-\mu}^{+1} p_i = 0$ and since at least one of the vectors $p_{-\mu},\dots,p_1$ is different from zero, we must have at least two of the vectors $p_{-\mu},\dots,p_1$ which are different from zero,and hence at least one of the vectors $p_{-\mu},\dots,p_0$ must be different from zero. Since $p_i \cdot x \geqslant 0$ whenever $\phi_i(0) + \text{grad } \phi_i(0)\cdot x<0$, it follows that for some $\lambda_i \leqslant 0$, we have $p_i = \lambda_i \text{ grad } \phi_i(0)$. We remark here that we may choose $\lambda_i=0$ for all $i \in \{-\mu,\dots,-1\}$ such that $\phi_i(0)<0$, since in that case we have $p_i=0$. We have $p_1 = -\Sigma_{i=-\mu}^{0} p_i$ and hence the relation $p_i \cdot x \geqslant 0$ for all $x \in \Omega_1$ may be written under the form $(\Sigma_{i=-\mu}^{0} p_i)\cdot x \leqslant 0$ for all $x \in \Omega_1$. The last inequality is equivalent to relation (iii) under the assumption $\hat{x}=0$. This concludes the proof of Theorem 4.2.

Remark 1. The inequality (iii) of Theorem 4.2 may be written under the form of a Maximum Principle:

$$(\Sigma_{i=-\mu}^{0} \lambda_i \text{ grad } \phi_i(\hat{x}))\cdot\hat{x} \geqslant (\Sigma_{i=-\mu}^{0} \lambda_i \text{ grad } \phi_i(\hat{x}))\cdot x \text{ for all } x \in \Omega_1.$$

Remark 2. If the point $\hat{x}$ is an interior point of Ω_1, then the inequality (iii) becomes

$$(iii)^* \quad \Sigma_{i=-\mu}^{0} \lambda_i \text{grad } \phi_i(\hat{x}) = 0.$$

This will always be the case for the problems where $S_1=\Omega_1=R^n$. This last form of Theorem 4.2 is known as Fritz John Theorem [3].

5. CONSTRAINT QUALIFICATIONS

We remark here that Theorem 4.2 contains no information about λ_0 besides the fact that $\lambda_0 \leq 0$. If we would know that $\lambda_0 < 0$, then we could multiply the entire vector λ by the positive number $-1/\lambda_0$ and we would obtain the same type of necessary conditions with some vector λ^* for which we would have $\lambda_0^* = -1$. A great variety of conditions (Constraint Qualifications) can be imposed on the problem which would allow us to guarantee that there exists some vector λ with $\lambda_0 < 0$. One of the major shortcomings of the traditional presentation of necessary conditions in the mathematical programming literature is, in my opinion, that the concept and the choice of those Constraint Qualifications influence the entire development of the theory of necessary conditions instead of being introduced at the last minute and used only to prove a variety of, practically important but theoretically easy, corollaries to Theorem 4.2. For example, a very general Constraint Qualification for the problem of Section 4 is to assume that $\overline{\Omega}_{-\mu} \cap \overline{\Omega}_{-\mu+1} \cap \cdots \cap \overline{\Omega}_{-1} \cap \Omega_1$ is a simplicial[1] convex approximation to $S_{-\mu} \cap S_{-\mu+1} \cap \cdots \cap S_{-1} \cap S_{+1}$. In the case $\Omega_1 = S_1 = R^n$, this Constraint Qualification is known as the Abadie Sequential Constraint Qualification. (See the remark following Definition 3.3).

6. LIMITATIONS OF THE METHOD OF DUBOVITSKII AND MILYUTIN

The method of Dubovitskii-Milyutin is not well adapted to problems with equality constraints. I shall examplify those difficulties by considering the following optimization problem in the plane R^2: minimize $\phi_0(x_1,x_2) \equiv x_1$ subject to the constraints $\phi_1(x_1,x_2) \equiv x_2 = 0$ and $\phi_2(x_1,x_2) \equiv x_2 - x_1^2 = 0$. The point $(\hat{x}_1,\hat{x}_2) = (0,0)$ is the obvious optimal solution of that problem. The sets $S_0 = \{(x_1,x_2) : \phi_0(x_1,x_2) < \phi_0(\hat{x}_1,\hat{x}_2)\} = \{(x_1,x_2) : x_1 < 0\}$, $S_1 = \{(x_1,x_2) : \phi_1(x_1,x_2) = 0\} = \{(x_1,x_2) : x_2 = 0\}$ and $S_2 = \{(x_1,x_2) : \phi_2(x_1,x_2) = 0\} = \{(x_1,x_2) : x_2 = x_1^2\}$ have no point in common, but the sets Ω_0, Ω_1 and Ω_2 have points in common (here, Ω_0 is the interior convex approximation to S_0, and Ω_i is the tangent convex approximation to S_i for i=1 and 2). Indeed, $\Omega_0 = \{(x_1,x_2) : x_1 < 0\}$ and $\Omega_1 = \Omega_2 = \{(x_1,x_2) : x_2 = 0\}$ and we have $\bigcap_{i=0,1,2} \Omega_i = \{(x_1,x_2) : x_1 < 0, x_2 = 0\} \neq \Phi$. Of course such "accidents" could be ruled out by conditions resembling some Constraint Qualifications. In the simple example given above for instance, we could require that the set of gradients of the equality constraints be linearly independent at the optimal point. If operator constraints are present in the problem, the situation would still be more complex and one would need further types of Constraint Qualifications. In the next section, I will present a theory of necessary conditions for optimization problems with equality and operator constraints (and also inequality constraints, but they never present any difficulties) which will be independent of any sort of constraint qualifications.

7. THE CASE OF INEQUALITY, EQUALITY AND OPERATOR CONSTRAINTS

The central part of this section is the following result.

Theorem 7.1. If $I=\{-\mu,\dots,m+1\}$ and if $\{S_i:i\in I\}$ and $\{\Omega_i:i\in I\}$ are families of subsets of R^n such that (i) $\bigcap_{i\in I}S_i=\Phi$, (ii) for $i=-\mu,\dots,0$, the set Ω_i is an interior convex approximation to the set S_i, (iii) for $i=1,\dots,m$, the set Ω_i is a tangent convex approximation to the set S_i, and (iv) Ω_{m+1} is an $(m+1)$-convex approximation to the set S_{m+1}, then the family $\{\Omega_i:i\in I\}$ is separated.

We remark immediately that in the case $m=0$, Theorem 7.1 coincides with Theorem 4.1. From the counterexample given in Section 6, we know that under the assumptions of Theorem 7.1, it would be incorrect to say (as in Theorem 4.1) that $\bigcap_{i\in I} S_i=\Phi$ implies that $\bigcap_{i\in I}\Omega_i=\Phi$,but we can still assert that the family $\{\Omega_i:i\in I\}$ is separated and this last statement is all that is needed to obtained appropriate necessary conditions. The proof of Theorem 7.1, given in Halkin [2], makes a critical use of Brouwer Fixed Point Theorem.

Let us now assume that we are faced with the following optimization problem: given a subset S of R^n and functions $\phi_{-\mu},\dots,\phi_m$ defined over R^n, find an element $\hat{x}\in R^n$ which minimizes $\phi_0(\hat{x})$ subject to

(α) the inequality constraints $\phi_i(\hat{x})\leq 0$ for $i=-\mu,\dots,-1$,

(β) the equality constraints $\phi_i(\hat{x})=0$ for $i=1,\dots,m$,

(γ) the operator constraint $\hat{x}\in S$.

The optimality of such element $\hat{x}$ can be expressed by writing that $\bigcap_{i=-\mu}^{m+1}S_i=\Phi$, where the sets $S_{-\mu},\dots,S_{m+1}$ are defined by

$S_i=\{x:x\in R^n,\phi_i(x)\leq 0\}$ for $i=-\mu,\dots,-1$,

$S_0=\{x:x\in R^n,\phi_0(x)<\phi_0(\hat{x})\}$,

$S_i=\{x:x\in R^n,\phi_i(x)=0\}$ for $i=1,\dots,m$

and $S_{m+1}=S$.

Let us assume that the functions $\phi_{-\mu},\dots,\phi_m$ are differentiable at $\hat{x}$ and that the functions $\phi_1,\dots,\phi_m$ are continuous in a neighborhood of $\hat{x}$. We then define convex sets $\Omega_{-\mu},\ldots,\Omega_m$ by the relations

$$\Omega_i = \{x : x \in R^n, \phi_i(\hat{x}) + \text{grad } \phi_i(\hat{x})\cdot(x-\hat{x})<0\} \text{ if } i=-\mu,\ldots,-1,$$

$$\Omega_0 = \{x : x \in R^n, \text{grad } \phi_0(\hat{x})\cdot(x-\hat{x})<0\}$$

and

$$\Omega_i = \{x : x \in R^n, \text{ grad } \phi_i(\hat{x})\cdot(x-\hat{x})=0\} \quad \text{for } i=1,\ldots,m.$$

As was mentioned in Section 3, the sets $\Omega_{-\mu},\ldots,\Omega_0$ are interior convex approximations around the point $\hat{x}$ to the sets $S_{-\mu},\ldots,S_0$ and the sets $\Omega_1,\ldots,\Omega_m$ are tangent convex approximations around the point $\hat{x}$ to the sets $S_1,\ldots,S_m$. Let us assume that we are given a set Ω_{m+1} which is a simplicial m+1 convex approximation around the point $\hat{x}$ to the set $S_{m+1}=S$. From Theorem 7.1, we know that the sets $\Omega_{-\mu},\ldots,\Omega_{m+1}$ will be separated. If we translate this last result in terms of the functions $\phi_{-\mu},\ldots,\phi_m$ and their gradients at the point $\hat{x}$, then, by an argument similar to the argument followed in the proof of Theorem 4.2, we obtain

Theorem 7.2. If $\hat{x}$ is an optimal solution of the given problem, then there exist numbers $\lambda_{-\mu},\ldots,\lambda_m$, not all zero, such that

(i) $\lambda_i \leq 0$ for $i=-\mu,\ldots,0$;

(ii) $\lambda_i\phi_i(\hat{x})=0$ for $i=-\mu,\ldots,-1$;

(iii) $\Sigma_{i=-\mu,\ldots,m} \lambda_i \text{grad } \phi_i(\hat{x})\cdot(x-\hat{x}) \leq 0$ for all $x \in \Omega_{m+1}$.

We conclude by making two remarks similar to the remarks made at the end of Section 4.

Remark 1. The inequality (iii) of Theorem 7.2 may be written under the form a Maximum Principle:

$$\Sigma_{i=-\mu,\ldots,m} \lambda_i \text{ grad } \phi_i(\hat{x})\cdot\hat{x} \geq \Sigma_{i=-\mu,\ldots,m} \lambda_i \text{grad } \phi_i(\hat{x})\cdot x \quad \text{for all } x \in \Omega_1.$$

Remark 2. If the point $\hat{x}$ is an interior point of Ω, then the inequality (iii) becomes

$$\Sigma_{i=-\mu,\ldots,m} \lambda_i \text{ grad } \phi_i(\hat{x}) = 0.$$

This will always be the case for the problems where $S_1=\Omega_1=R^n$.

REFERENCES

[1] Dubovitskii,A.Ya, and Milyutin,A.A., Extremum Problems in the Presnece of Restrictions, U.S.S.R. Computational Mathematics and Mathematical Statistics, 5, 1965, 1-79.

[2] Halkin,H., A Satisfactory Treatment of Equality and Operator Constraints in the Dubovitskii-Milyutin Optimization Formalism, Journal of Optimization Theory and Applications, 6, 1970, 138-149.

[3] John,F., Extremum Problems with Inequalities as Subsidiary Conditions, in "Studies and Essays:Courant Anniversary Volume",(K.O.Friedricks,O.E.Neugebauer,and J.J. Stoker,(eds.)),pp.187-204,Interscience Publishers,New York,1948.

[4] Abadie,J., On the Kuhn-Tucker Theorem, in "Nonlinear Programming",J.Abadie(ed.), pp.19-36,North-Holland,1967.

[5] Halkin,H., Optimal Control as Programming in Infinite Dimensional Spaces, in "C.I.M.E.:Calculus of Variations,Classical and Modern",pp.179-192,Eidizioni Cremonese,Roma,1966.

[6] Halkin,H. and Neustadt,L.W., Control as Programming in General Normed Linear Spaces, Lecture Notes in Operations Research and Mathematical Economics,Springer Verlag,11,1969,23-40.

OPTIMUM DESIGN OF STRUCTURES
THROUGH VARIATIONAL PRINCIPLES

by RICHARD T. SHIELD

Department of Theoretical and Applied Mechanics
University of Illinois, Urbana, U.S.A.

1. INTRODUCTION

The application of the calculus of variations to the design of structures for minimum volume leads to necessary conditions for the structural volume to be stationary, and a local or global minimum is not guaranteed. However, if an appropriate variational principle applies for the class of structures under consideration, design criteria can be established which lead to structures of minimum volume. A direct design procedure was first given by Michell [1] for framed structures composed of a material of limited strength. For perfectly-plastic structures, direct design procedures were introduced by Drucker and Shield [2,3,4] and here the upper bound theorem of limit analysis provided the appropriate variational principle. For elastic structures, variational principles can provide direct design methods for design for a given stiffness, for a given buckling load or for given fundamental frequency of vibration (see Prager and Taylor [5] and Shield and Prager [6]). The major part of this paper surveys the direct design procedures which have been developed through the use of variational principles.

Section 2 describes the procedures for minimum-volume design of structures of perfectly-plastic materials which are required to carry a given set of loads. Section 3 discusses uniform strength designs in which the structural material is required to be stressed within a certain range under a given system of loads. The stress range may be chosen to ensure that the stresses remain in the elastic range, for example, or to ensure that an appreciable amount of creep will not occur. Section 4 treats elastic design for a given stiffness in order to illustrate the design procedures for elastic structures. Minimum-volume framed structures of material of limited strength in tension and compression are considered in Section 5. The Michell design method fails when kinematic constraints are present (except when the tensile and compressive strengths are equal) but an alternative approach [7] does not have this limitation. An example is given to show that minimum-volume frames are not necessarily unique, and some new plane structures of the Michell type are described, including the layout for pure bending.

This paper is not intended to provide a comprehensive survey of the literature on optimum design. The reader will find additional references in [7, 8, 9, 10].

2. PLASTIC DESIGN OF STRUCTURES

In this section we discuss the optimum design of structures composed of perfectly-plastic materials. A restricted formulation of the problem is indicated in Figure 1. It has the advantage of ensuring that the structure will consist of the usual structural elements, such as frames, plates and shells, and the limitations also increase the chances of determining the optimum solution. We suppose that the structure is to have a prescribed middle surface A. The loading is prescribed and is distributed over A and its boundary. At supports either the components of displacement and rotation are prescribed to be zero or the corresponding components of edge traction and moment are given. For a solid shell, problem (i), the structure is formed by placing a certain thickness h of a given material at points of the middle surface. For a sandwich type structure, problem (ii), we suppose that the shell has a core of prescribed thickness H. The core carries shear force only, and bending moments and force resultants are carried by membrane stresses in two thin identical face sheets of thickness h. In both cases we wish to design the shell, that is choose h, so that the shell is just at collapse under the given loading and is optimum for a given criterion. Here we design so that the volume

$$V = \int h \, dA$$

is minimized but the methods extend readily to minimization of the functional

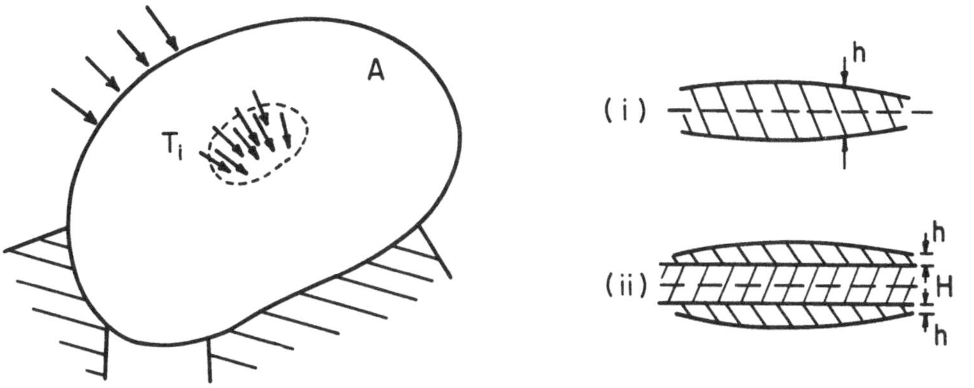

Figure 1. Solid and sandwich shells

$$\int h\, f(X)\, dA,$$

where f (X) is a non-negative function of position over A. The extension allows for minimum-weight design when different materials are to be used in different parts of A, and also admits design for minimum moment of inertia about an axis.

The generalized stresses, such as bending moments and stress resultants across a section, will be denoted by Q_1, Q_2, .., Q_N or simply Q_n. Yielding can occur in the shell when the stresses Q_n lie on the yield surface

$$F\,(Q_n;\, h) = 0 \tag{1}$$

in N-dimensional stress space. For stress states Q_n represented by points on the yield surface, purely plastic strain rates are possible; and we use q_n (n = 1, 2, .., N) to denote the generalized strain rates, such as rates of curvature change and extension of the middle surface. The vector representing q_n is normal to the yield surface at regular points, while at singular points the vector lies between adjacent normals. For a convex yield surface, the plastic rate D_A of dissipation of energy per unit area of the middle surface is then uniquely determined by the values of q_n,

$$D_A = D_A\,(q_n;\, h) = Q_1\, q_1 + Q_2\, q_2 + \,..\, + Q_N\, q_N. \tag{2}$$

For the solid shell, D_A is quadratic in h in general; while for the sandwich shell, D_A is directly proportional to h. Shear forces have little influence on yielding, even for highly localized loading [11], and they are not included in the generalized stresses Q_n.

The theorems of limit analysis [12, 13] can be used to provide information about the volume of a design which can carry the loads. The appropriate theorems are the following:

Lower-bound theorem. If the applied loads can be carried by an equilibrium distribution of moments and force resultants Q_n in the shell which are at or below yield, the loading is at or below the collapse loading.

Upper-bound theorem. If the applied loads are such that a deformation of the shell can be found for which the rate at which the applied loads do work exceeds the rate of internal energy dissipation, the loading is above collapse.

The lower-bound theorem can be used to determine upper bounds on the volume of the minimum-volume design for given loads [3]. If a stress distribution Q_n over A is in equilibrium with the applied loads, we can choose the thickness h_s so that the yield condition (1) is satisfied everywhere on A. Since the design h_s is then a permissible design, the minimum volume V_m must be less than or equal to V_s,

$$V_m \leq V_s = \int h_s \, dA.$$

The upper-bound theorem can be used to provide lower bounds for V_m, as in [3], but the theorem also leads to a direct design procedure [4]. For a shell h_s which is at or below collapse under the applied loads and for any kinematically admissible velocity field u_i $(i = 1, 2, 3)$ we have

$$\int D_A (q_n; h_s) \, dA - \int T_i \, u_i \, dA \geq 0, \tag{3}$$

for otherwise the use of the upper-bound theorem with the deformation u_i would predict that the loading exceeds collapse. In (3), T_i are the components of the applied loads, the repeated index i implies summation over 1, 2, 3, and the integral of $T_i \, u_i$ is to include the rate of work at the edge of the shell. Inequality (3) is a variational principle for the permissible design h_s; equality holds in (3) only when u_i is a collapse mode for h_s. The application of this principle to the determination of a minimum-volume design is much more direct than the use of the calculus of variations. Suppose that a design h_c for a solid shell, problem (i), is just at collapse under the loads in a collapse mode u_i, and suppose that a neighboring design $h_s = h_c + \delta h$ is also a permissible design. If we neglect second order terms, the dissipation rate for the shell h_s in the deformation mode u_i is

$$D_A (q_n; h_c) + \delta h \frac{\partial D_A}{\partial h},$$

assuming that D_A is continuously differentiable in h. Applying the variational principle (3) to the design h_s, we conclude that

$$\int \delta h \frac{\partial D_A}{\partial h} dA \geq 0$$

because equality holds in (3) for the design h_c. It now follows that if the design h_c is such that

$$\frac{\partial}{\partial h} D_A (q_n; h) = \text{constant} \tag{4}$$

over A then

$$\int \delta h dA \geq 0, \tag{5}$$

so that the design h_c provides a relative minimum for the volume of permissible designs. Thus, assuming that the neglect of the second-order terms is permissible, the variational principle leads to a direct design procedure. Mroz [14] has given an example in which the application of (4) leads only to a stationary value for the volume.

The case when D_A is directly proportional to h, as for the ideal sandwich shell of problem (ii), is simpler and a stronger result is possible. We again suppose that the design h_c is at collapse in a deformation mode u_i, and we use the mode u_i in the variational principle (3) for another permissible design h_s. We obtain

$$\int D_A (q_n; h_s) dA \geq \int T_i u_i dA = \int D_A (q_n; h_c) dA.$$

Since

$$D_A (q_n; h_s) = D_A (q_n; h_c) \frac{h_s}{h_c},$$

we can conclude that if the design h_c is such that

$$\frac{D_A (q_n; h)}{h} = \text{constant} \tag{6}$$

over A, then

$$\int h_s \, dA \geq \int h_c \, dA.$$

Not only does the condition (6) lead to an absolute minimum for the design volume but the condition (6) is much easier to use than condition (4) because (6) does not involve the design thickness directly.

In order to illustrate the use of these design methods we consider the minimum-volume design of a circular plate with a built-in edge under uniform pressure loading on its upper face (for problems involving other symmetric and non-symmetric pressure distributions see [15, 16]). When the Tresca yield condition is assumed, the yield condition on the radial bending moment M and the circumferential bending moment N is the familiar hexagon in (M, N) space,

$$\max (|M| , |N| , |M - N|) = M_0,$$

in which $M_0 = \sigma_0 Hh$ for the sandwich plate and $M_0 = \frac{1}{4} \sigma_0 h^2$ for the solid plate. The curvature rates κ, λ in the radial and circumferential directions are derived from the downward deflection rate w of the middle surface through

$$\kappa = - \frac{d^2 w}{dr^2}, \qquad \lambda = - \frac{1}{r} \frac{dw}{dr},$$

where r measures distance from the center.

For the sandwich plate, the design criterion (6) does not involve the design thickness h, and it is readily found that (6) can only be satisfied for a finite range of r when either

$$\text{(i)} \quad M = N = \pm M_0, \qquad \text{or} \qquad \text{(ii)} \quad M = \pm M_0, \quad N = 0.$$

For these regimes, condition (6) reduces to

$$\text{(i)} \quad \kappa + \lambda = \pm \alpha, \qquad \text{or} \qquad \text{(ii)} \quad \kappa = \pm \alpha,$$

where α is a constant when the core thickness H is constant. The curvature rates κ, λ must have the same sign in regime (i) while in regime (ii) they are of opposite sign and $|\kappa| \geq |\lambda|$. For a plate with a built-in edge, regime (i) with the positive sign will apply in a central region $r \leq a$ and regime (ii) with the negative sign will apply in the remaining portion $a \leq r \leq R$. At the built-in edge w and dw/dr are zero and dw/dr is zero at the center. In order to have w and dw/dr continuous at the junction $r = a$, it is found that we

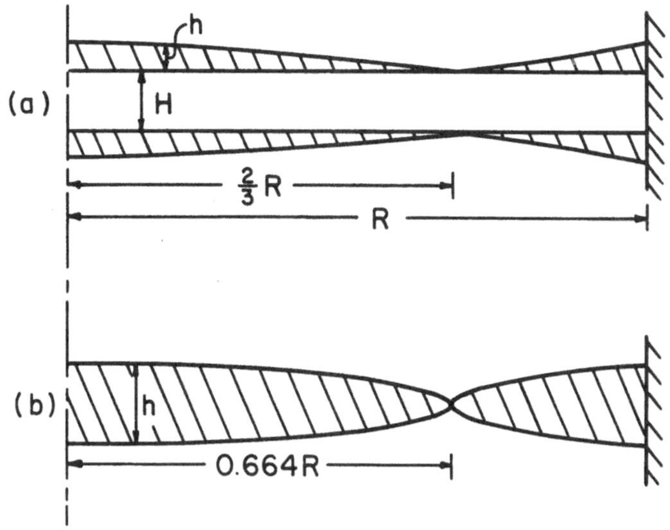

Figure 2. Minimum-volume designs for built-in circular plate under uniform pressure (a) sandwich plate (b) solid plate

must have $a = 2R/3$, and at this radius M must vanish because κ changes sign. With $M = 0$ at $r = 2R/3$, the moment distribution in the two regions of the plate, and hence the thickness $h = |M| / \sigma_0 H$, can now be found from equilibrium. The design is indicated in Figure 2 (a).

For the solid plate, we again assume that M and N have the fully plastic value M_o for $r \leq a$ and that N is zero for $a \leq r \leq R$, with $M = 0$ at $r = a$. From equilibrium the moment distribution and therefore the thickness $h(r)$,

$$h = \left\{ 4 \; |M| / \sigma_0 \right\}^{\frac{1}{2}},$$

can be found for a general value of a. In order to satisfy the design criterion (4) we must have

$$\frac{d^2 w}{dr^2} + \frac{1}{r} \frac{dw}{dr} = - \frac{\alpha}{h} \quad \text{and} \quad \frac{d^2 w}{dr^2} = \frac{\alpha}{h}$$

in the inner and outer regions, respectively. The continuity of dw/dr at $r = a$ leads to (see [15])

$$\int_0^a \frac{r\,dr}{h(r)} = a \int_a^R \frac{dr}{h(r)} ,$$

and this equation serves to determine the junction radius a. For uniform pressure loading $a = 0.664\,R$ and the design is as indicated in Figure 2 (b).

Minimum-volume design for other one-dimensional situations, such as symmetrically loaded circular cylindrical shells of sandwich type [4], is also relatively straight-forward, but design problems which are two-dimensional can be much more difficult to treat. So far designs for non-circular plates with built-in edges have only been obtained by an inverse method [17].

The design procedure can be modified to include body forces (such as weight) which act only when material is present (see [4]). Also the procedure has been extended to the design of multi-purpose structures which are to support different systems of loads at different times [18], and to the quasi-static design of structures under moving loads [19].

3. UNIFORM STRENGTH DESIGNS

The methods for plastic design have been extended [7] to materials which are not perfectly plastic, so that design limitations other than plastic collapse are involved. For example, we may use a work-hardening material and in order to minimize the possibility of fracture we may wish to design the shell so that it remains elastic everywhere. For another material, we may wish to keep the stresses below the level at which an appreciable amount of creep will occur. In both cases there will be a limiting surface in stress space to restrict the stress states in a section of the shell. We shall say that a design is a uniform strength design for the given loading if the stresses everywhere in the shell are on the limiting surface. We seek the uniform strength design which has least volume.

Figure 3 indicates a limiting surface in generalized stress space. We suppose that the surface is given by

$$L\,(Q_n;\,h) = 0, \tag{7}$$

where L is a known function, and we assume that the surface is convex. In purely elastic design, for example, the limiting surface is determined by those values Q_n for which the yield limit is reached in the outer fibers of the shell.

Consider an infinitesimal virtual deformation of the shell defined by middle surface displacements v_i and associated generalized strains e_n. We shall say that a virtual defor-

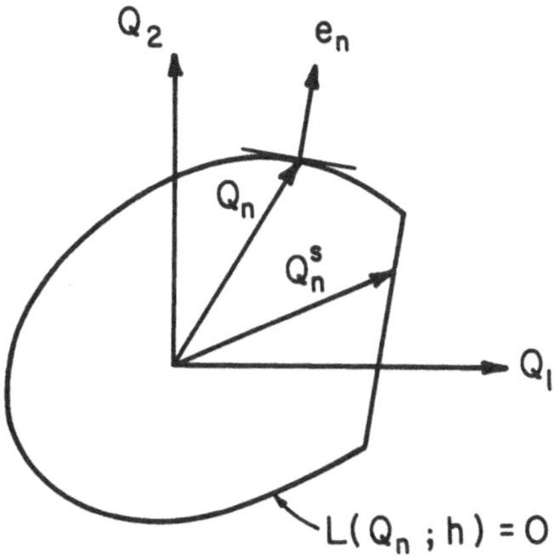

Figure 3. Limiting surface in generalized stress space

mation with strains e_n is compatible with a limiting stress state Q_n if the vector representing e_n is normal to the limiting surface at the stress point Q_n, Figure 3. At a singular point of the limiting surface the strain vectors representing compatible deformations lie in the fan bounded by adjacent normals. For a given convex limiting surface, the virtual work W_A in a compatible virtual deformation is then uniquely determined by the virtual strains e_n,

$$W_A = Q_n e_n = W_A (e_m; h),$$

in which the repeated index n implies summation over 1, 2, ..., N. Moreover, for any other stress state Q_n^s inside or on the limiting surface we have

$$Q_n^s e_n \le W_A (e_m; h) \tag{8}$$

with equality only if e_n are also compatible with Q_n^s.

The approach to minimum-volume design for uniform strength is similar to that for plastic design, and we shall treat the case when W_A is directly proportional to the design thickness h. If the shell h_c has limiting stresses under the loads and if there is an admissible compatible virtual deformation v_i, then by virtual work we have

$$\int T_i v_i \, dA = \int W_A (e_n; h_c) \, dA. \tag{9}$$

For any other shell h_s with stresses Q_n^s at or below the limiting values we can use (8) and virtual work to deduce that

$$\int T_i v_i \, dA = \int Q_n^s e_n \, dA \leq \int W_A (e_m; h_s) \, dA,$$

where v_i is any admissible virtual displacement. We therefore have the variational principle

$$\int W_A (e_n; h_s) \, dA - \int T_i v_i \, dA \geq 0 \tag{10}$$

for permissible designs and equality holds only when h_s is a uniform strength design and v_i is compatible with the stresses Q_n^s in the shell. We now use the compatible virtual deformation v_i for the design h_c in the variational principle (10) and with (9) we derive

$$\int W_A (e_n; h_s) \, dA \geq \int W_A (e_n; h_c) \, dA.$$

Because W_A is proportional to h, we see that if the design h_c is such that

$$\frac{W_A (e_n; h)}{h} = \text{constant}$$

over A, then the volume of h_c will be an absolute minimum for all permissible designs.

For the solid shell the design criterion is

$$\frac{\partial}{\partial h} W_A (e_n; h) = \text{constant}$$

over the shell.

Uniform strength designs have been discussed by Save [20].

4. ELASTIC DESIGN FOR GIVEN STIFFNESS

Direct design methods can be developed in the same way for other problems of optimum design provided that a suitable variational principle holds for the structure under investigation. This can be the case in the minimum-volume design of an elastic structure which is to have a given stiffness under a given set of loads (or, equivalently, elastic design for maximum stiffness with a given volume of material). Other examples are minimum-volume design for a given buckling load or for a given fundamental frequency of vibration. Techniques for design problems such as these that have been developed in a unified way by Prager and Taylor [5]. Here we outline the procedure in the case of elastic design for a given stiffness.

For an elastic shell there is a strain-energy function E_A, per unit area of the middle surface, which is uniquely determined by the generalized strains q_n derived from middle surface displacements u_i. The strain energy also depends on the design thickness h so that we write it as $E_A(q_n; h)$. The potential energy U is defined as

$$U\{u^*; h\} = \int E_A(q_n^*; h)\, dA - \int T_i u_i^*\, dA,$$

where the integral of $T_i u_i$ represents all the virtual work of the prescribed loads including the edge loading and where u_i^* is a displacement field which satisfies any imposed displacement conditions. When E_A is a positive definite quadratic function of the strains, the Principle of Minimum Potential Energy holds. The principle states that the potential energy U is minimized by the actual displacements u_i produced by the loads,

$$U\{u^*; h\} \geq U\{u; h\}.$$

We now define the compliance of the shell for the given loads to be twice the total strain-energy of the shell and we note that

$$2\int E_A(q_n; h)\, dA = \int T_i u_i\, dA.$$

For two designs h and h_s with the same compliance, we have

$$\int E_A(q_n; h_s)\, dA \geq \int E_A(q_n^s; h_s)\, dA = \int E_A(q_n; h)\, dA, \tag{11}$$

where q_n^s are the strains for the design h_s. The inequality in (11) follows from the Principle of Minimum Potential Energy applied to the design h_s. When E_A is directly proportional to h, we see from (11) that in designing for a given compliance, the design with E_A/h constant will have least volume. For other types of shells the procedure would be to design so that $\partial E_A/\partial h$ is constant over the shell, and the design would provide a relative minimum for the volume of permissible designs.

As a simple example, suppose we have an elastic beam of length 2ℓ which is built-in at both ends and which has a transverse point load P at the center. We wish to design the beam so that the central deflection does not exceed δ and such that the beam has minimum volume. For a beam of the sandwich type, minimizing the volume is the same as minimizing the integral of the bending stiffness over the beam. If two beams with stiffnesses s and $\overline{s}$ have the same central deflection δ under the load, they have the same compliance $P\delta$ and in the same way that (11) was derived we can use the Principle of Minimum Potential Energy to get

$$\int \bar{s}\kappa^2\, dx \geq \int s\kappa^2\, dx,$$

where κ is the curvature of the design s under the load P and x measures distance from one end. We now see that the design s will have least volume if $|\kappa|$ is constant. In order to satisfy the constraints at the ends, the deflection with constant $|\kappa|$ must have inflection points at the quarter points $x = \ell/2,\ 3\ell/2$. Since the moment $M = s\kappa$ must vanish at the quarter points where κ changes sign, the moment distribution is now statically determinate and $M(x)$ and therefore $s(x)$ can be found.

The design procedure obtained from the Principle of Minimum Potential Energy applies for design with given compliance. However, the design criterion does not always coincide with the compliance. Thus if we have a distributed load over the built-in beam and we wish to limit the central deflection as before, the compliance will not be known in advance. Similarly, if we have an off-center point load P at the section $x = x_o$ and we wish to limit the maximum deflection of the beam, the compliance is $P u_o$, where u_o is the deflection at $x = x_o$ and is not necessarily the maximum deflection. These design problems can be approached by using a variational principle of a different type called the Principle of Stationary Mutual Potential Energy [6]. Let u_i and $\bar{u}_i$ be two middle surface displacement fields for a design of thickness h and let q_n, Q_n and $\bar{q}_n$, $\bar{Q}_n$, respectively be the associated generalized strains and stresses. We define the <u>mutual strain energy</u> through

$$E_A(q_n, \bar{q}_n; h) = \sum_1^N Q_n \bar{q}_n = \sum_1^N \bar{Q}_n q_n.$$

For two different sets T_i and $\bar{T}_i$ of applied loads, the <u>mutual potential energy</u> U_M is defined as

$$U_M\{u^*, \bar{u}^*; h\} = \int E_A(q_n^*, \bar{q}_n^*; h)\, dA - \int T_i \bar{u}_i^*\, dA - \int \bar{T}_i u_i^*\, dA,$$

where u_i^*, $\bar{u}_i^*$ are kinematically admissible displacement fields.

If u_i and $\bar{u}_i$ are the actual displacements that the loads T_i and $\bar{T}_i$, respectively, would induce in the shell, then

$$U_M\{u, \bar{u}; h\} = -\int T_i \bar{u}_i\, dA = -\int \bar{T}_i u_i\, dA. \tag{12}$$

With the use of the Principle of Virtual Work, it can now be shown (see [6] for details) that

$$U_M\{u^*, \bar{u}^*; h\} - U_M\{u, \bar{u}; h\} = \int E_A(q_n^* - q_n, \bar{q}_n^* - \bar{q}_n; h)\, dA. \tag{13}$$

If we apply (13) when u_i^* and $\bar{u}_i^*$ are neighboring displacements to the actual displacements u_i and $\bar{u}_i$, the right-hand side will be zero to first order. Thus $U_M\{u^*, \bar{u}^*; h\}$ is stationary at the values $u_i^* = u_i$, $\bar{u}_i^* = \bar{u}_i$, and this is the Principle of Stationary Mutual Potential Energy.

Suppose we wish to design a structure so that the transverse deflection at a particular point X_0 of the mid-surface is of amount δ under the loads T_i. We take the second system of loads $\bar{T}_i$ to be a single unit concentrated load $\bar{P}$ acting normal to the middle surface at the point X_0. From (12) we then see that the value of $U_M\{u, \bar{u}; h\}$ is $-\bar{P}\delta$, so that designs which satisfy the design criterion will have the same value for $U_M\{u, \bar{u}; h\}$. We can therefore use the Principle of Stationary Mutual Potential Energy in the same way as the Principle of Minimum Potential Energy was used in design for a given compliance. In this way we find that the design such that

$$\frac{\partial}{\partial h} E_A(q_n, \bar{q}_n; h) = \text{constant}$$

over the shell will provide a stationary value for the volume for designs which have transverse deflection of amount δ at the point X_0.

Applications to the minimum-volume design of beams for given deflections (or rotations) are described in [6]. Suppose we wish to design a beam of sandwich type and we require the deflection at the section $x = x_0$ to be of amount δ under a certain system of loads. Let s and s^* be the bending stiffnesses of two designs that satisfy the constraint on the deflection at x_0, and let u, u^* and $\bar{u}$, $\bar{u}^*$ be the corresponding deflections of these designs under the given loads and under a unit concentrated load $\bar{P}$ at x_0. From (12) we have

$$U_M\{u, \bar{u}; s\} = U_M\{u^*, \bar{u}^*; s^*\} = -\bar{P}\delta,$$

where we have identified the bending stiffnesses s and s^* with the design thicknesses h and h^*, as we may do for sandwich beams. The deflections u, $\bar{u}$ are kinematically admissible for the design s^* and if we apply (13) to this design we get

$$U_M\{u, \bar{u}; s^*\} - U_M\{u^*, \bar{u}^*; s^*\} = \int s^*(\kappa^* - \kappa)(\bar{\kappa}^* - \bar{\kappa})\, dx, \tag{14}$$

where κ, $\bar{\kappa}$, ... are the curvatures associated with the deflections u, $\bar{u}$, ... If we replace

$U_M\{u^*, \bar{u}^*; s^*\}$ by $U_M\{u, \bar{u}; s\}$ in (14) and use the definition of U_M we find that

$$\int (s^* - s)\,\kappa\,\bar{\kappa}\,dx = \int s^*\,(\kappa^* - \kappa)\,(\bar{\kappa}^* - \bar{\kappa})\,dx. \tag{15}$$

When s^* is a neighboring design to s, the right-hand side of (15) is zero to first-order and we see that

$$\kappa\,\bar{\kappa} = \text{constant} = c^2 \tag{16}$$

is a sufficient condition for the design s to provide a stationary value for the volume $\int s\,dx$. If $M = s\,\kappa$ and $\bar{M} = s\,\bar{\kappa}$ are the bending moments for the optimum design s under the two systems of loads then

$$s = (M\bar{M})^{\frac{1}{2}}/(\kappa\,\bar{\kappa})^{\frac{1}{2}} = \frac{1}{c}\,(M\bar{M})^{\frac{1}{2}}.$$

The constant c can be determined from $U\{u, \bar{u}; s\} = -\bar{P}\delta$ and we finally arrive at

$$s = \frac{(M\bar{M})^{\frac{1}{2}}}{\bar{P}\delta} \int (M\bar{M})^{\frac{1}{2}}\,dx. \tag{17}$$

For a statically determinate beam, the moment distributions M, $\bar{M}$ can be determined directly so that the optimum design (17) is readily found without calculation of deflections. Moreover, for a statically determinate beam it can be shown [6] that the design satisfying (16) actually furnishes an absolute minimum for the design volume. In this case the moment distributions M and $\bar{M}$ are independent of the stiffnesses, so that

$$M = s\,\kappa = s^*\,\kappa^*, \qquad \bar{M} = s\,\bar{\kappa} = s^*\,\bar{\kappa}^*.$$

These equations imply that

$$\kappa^* - \kappa = -\frac{(s^* - s)\,\kappa}{s^*}, \qquad \bar{\kappa}^* - \bar{\kappa} = -\frac{(s^* - s)\,\bar{\kappa}}{s^*}.$$

Substituting in (15) we obtain, with (16),

$$\int (s^* - s)\,dx = \int \frac{(s^* - s)^2}{s^*}\,dx \geq 0,$$

which shows that the design s satisfying (16) provides an absolute minimum for the design volume.

For a statically indeterminate beam, an extra step is required in order to arrive at a design that provides an absolute minimum for the volume. Consider, for example, a beam of length 2ℓ which is built-in at both ends and is loaded by a uniform pressure p along its length. We wish to restrict the deflection at the center $x = \ell$ to be of amount δ. For a beam built-in at both ends, $u''(x)$ must change sign at least twice for otherwise no deflection is possible; thus $M(x)$ will have at least two zeros. Assuming a symmetrical design, we suppose that the bending moment $M(x)$ is zero at $x = \ell \pm b$. If we now consider only designs for which the stiffness vanishes at $x = \ell \pm b$, we have a statically determinate beam and we can determine the design (17) which has least volume in this class of designs. We can now choose b so that the volume will have the least value for all possible designs, and this value of b is found to be $\ell/2.01$.

When the loading is not symmetric, the maximum deflection may be off center. Suppose, for example, that we have a simply supported beam of length 2ℓ under a system of loads which produces a bending moment $M(x)$. We wish to limit the maximum deflection to an amount δ. Let $\overline{M}(x)$ be the bending moment distribution caused by a unit point load $\overline{P}$ at $x = b$. The design (17) will then be optimum for a deflection of amount δ at $x = b$. We can ensure that the section $x = b$ will have the greatest deflection if we choose b so that $u'(x)$ is zero at $x = b$. In order to determine b, we note that if $u'(b) = 0$ and $u = 0$ at the ends, then

$$u(b) = -\int_0^b \int_y^b u''(x)\,dx\,dy = -\int_b^{2\ell} \int_b^y u''(x)\,dx\,dy,$$

and this implies that

$$\int_0^b x\kappa\,dx = \int_b^{2\ell} (2\ell - x)\,\kappa\,dx. \qquad (18)$$

Because $\kappa = M/s = c\,(M/\overline{M})^{\frac{1}{2}}$, we can write (18) as

$$\int_0^b x\,(M/\overline{M})^{\frac{1}{2}}\,dx = \int_0^{2\ell} (2\ell - x)\,(M/\overline{M})^{\frac{1}{2}}\,dx,$$

and this equation serves to determine b. To give an example, when the beam is loaded by a point load P at $x = a$, the value of b varies from ℓ to 1.11ℓ as a varies from ℓ to 2ℓ.

The procedures described here for elastic design can be extended to design with two or more constraints on deflection or rotation under a single system of loads [6] and to the design of multi-purpose structures [6, 21].

5. MICHELL STRUCTURES

In formulating the problem of optimum design in Section 2, we assumed that the type of the structure and the layout, that is the middle surface A, were specified. A less restrictive formulation merely specifies the region in which the given material can be placed and leaves the type and layout of the structure to be determined. In 1904 Michell published his paper [1] on the minimum-volume design of framed structures. He specified that the structure should consist of tie-bars in tension and struts in compression, but the layout of the structure was not specified. The material to be used allows a maximum tensile stress σ_t and a maximum compressive stress σ_c, and for a design which carries the prescribed loads, the minimum volume allowable is

$$V = \Sigma \ell_t f_t/\sigma_t + \Sigma \ell_c f_c/\sigma_c. \tag{19}$$

Here f_t is the tension in any tie-bar of length ℓ_t and f_c is the thrust in any strut of length ℓ_c. Michell showed that a framed structure will be of minimum volume if there is a virtual small deformation of the space such that each tie-bar suffers an extensional strain of amount e and each strut suffers a compressive strain of amount e and no linear element of space suffers a strain numerically greater than e, where e is a constant. Note that the actual deformation of the minimum-volume frame under the loads involves extensional and compressive strains of amounts σ_t/E and σ_c/E, respectively, along the frame elements, where E is Young's modulus.

In the proof of his results, Michell used a theorem due to Maxwell. By imposing a uniform dilatation on the whole of space, Maxwell showed that for all structures under the same system of applied loads

$$\Sigma \ell_t f_t - \Sigma \ell_c f_c = \text{constant}.$$

(The constant is $\Sigma \underline{F} . \underline{r}$, where $\underline{F}$ is an applied load at a point with position vector $\underline{r}$.) However, Maxwell's theorem does not apply to structures with kinematic constraints imposed by support conditions because the reactions at the supports can vary with the structure. An exception is a structure with one fixed point but in this case the reaction at the support is determined by overall equilibrium. Because Maxwell's theorem is essential to Michell's proof when $\sigma_t \neq \sigma_c$, the design procedure of Michell will not be valid in general when kinematic constraints are imposed. This limitation on the use of Michell's method does not appear to have been mentioned explicitly in the literature.

An alternative approach, which does not have the limitation of the Michell method, has been given by Shield [7]. The procedure is to design a frame compatible with a virtual small deformation in which the principal strains are of magnitude e/σ_t if extensional and of magnitude e/σ_c if compressive, the directions of frame elements coinciding with the principal directions of strain as before. The virtual deformation must satisfy any imposed kinematic constraints. The proof that the procedure leads to a minimum-volume frame is straightforward and it makes direct use of the Principle of Virtual Work as in the method of Section 3 for uniform strength designs. The proof has been repeated by Hemp [22] and by Hegemier and Prager [23] for the case $\sigma_t = \sigma_c$ (when the Michell method and the alternative method become identical).

Michell [1] supplied some examples of minimum-volume framed structures and other examples are given in [22,24,25,26]. Cox [27] has shown that a Michell structure has greater stiffness under the loads than any other structure which is stressed to the limiting values σ_t and σ_c. More recently, Hegemier and Prager [23] have shown that an elastic frame with a specified stiffness (i.e. compliance) has least volume when it has the layout of a Michell structure, and this holds also for frames designed for a given stiffness in stationary creep or for a given fundamental frequency of vibration. In the following we give an example to show that minimum-volume frames are not necessarily unique, and we describe some new additions to the list of Michell structures.

The diagram at the top of Figure 4 indicates the layout given by Michell [1] for a single force applied at the midpoint C of the line AB and balanced by equal parallel forces at A and B. The struts AD, EB and the curved bar DE carry a uniform compressive force and a quadrantal fan of tie-bars from C to DE maintains the equilibrium of the curved bar. The layout is symmetrical about AB with tie-bars replacing struts and vice-versa. The virtual deformation with principal strains $\pm$ e associated with the layout can be adjusted so that the

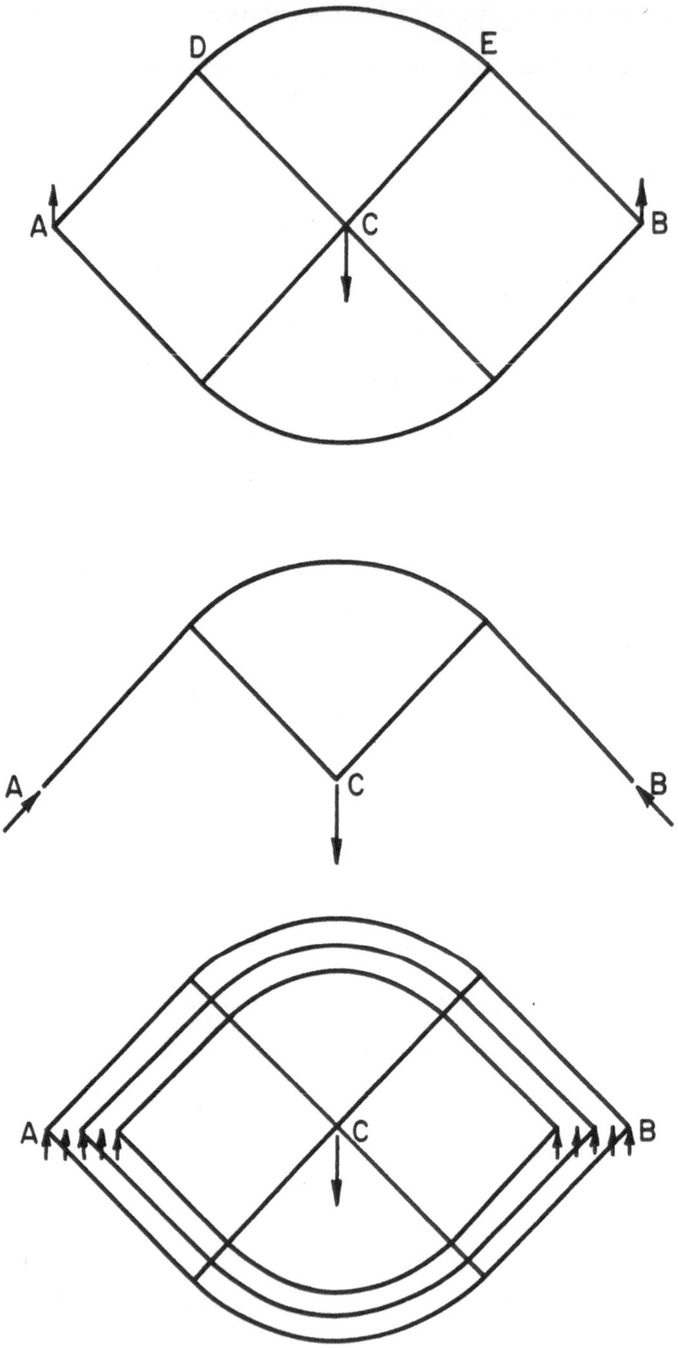

Figure 4. Load at C supported at A, B

displacement is zero at points A and B. If we assume $\sigma_t = \sigma_c$, we can use this virtual deformation for the case when we have the same force at C, but now A and B are fixed points of support. The optimum structure has the same volume as the structure with specified parallel forces at A, B, but the optimum design is not unique. For example, the load at C can be carried by a frame entirely above AB as indicated in the middle diagram of Figure 4. An infinity of optimum designs results from arbitrarily assigning a fraction of the load at C to be carried by a structure above the line AB and the remainder by a structure below the line AB. We note that if we had specified that the load at C be carried by a beam with center-line AB and built-in at A and B, the optimum design would have bending moments at A and B. The Michell structure has no moments at the fixed points A, B.

The minimum-volume design indicated at the bottom of Figure 4 uses the same virtual deformation with principal strains $\pm$ e, but now it is specified that distributed loads at A and B balance the load at C.

Figure 5 shows the optimum layout for pure bending. A bending moment at the point A is to be transmitted to the point B by a framed structure of minimum volume, composed of a material of limited strength (or the structure has an assigned bending stiffness). In the circular regions around the points A and B, the tie-bars and struts follow logarithmic spirals. The spiral regions are connected by a strut GH in compression at the top and a tie-bar carrying the same force at the bottom of the structure. The associated virtual deformation with principal strains $\pm$ e is, apart from a rigid displacement, purely circumferential in the circular regions. The regions between the larger circles and the straight-line boundaries (such as GC, CH) of the upper and lower quadrants which meet at C move as rigid bodies. In the quadrants meeting at C, the principal strain directions are vertical and horizontal, and the quadrants deform like a plastic hinge in a beam in pure bending. The total volume of material required is

$$M \left[1 + 2 \ln \frac{a}{\sqrt{2}\, r_0}\right]\left[\frac{1}{\sigma_t} + \frac{1}{\sigma_c}\right] .$$

Here M is the moment applied at A and B, a is the length of AC or AB and r_0 is the radius of small circles at A and B over which the forces equivalent to the moments M are distributed.

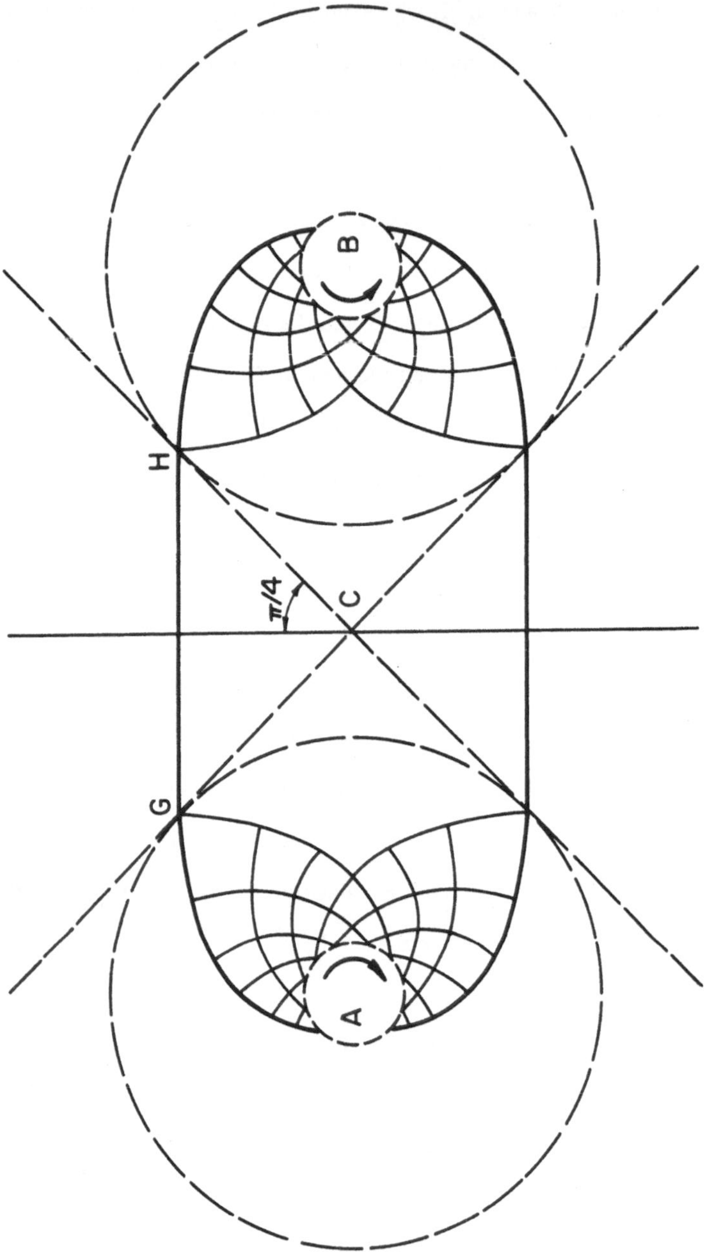

Figure 5. Layout for pure bending

Figure 6 indicates the optimum layout when a downward force P is added at the central point C and upward parallel forces P/2 are added at the points A and B. The moment M applied at A and B and the force P are related to the angle 2α of the fan regions through

$$4M/Pa = \cot\alpha - 1.$$

As the ratio P/M increases, the angle α tends to $\pi/4$ and the structure approaches the Michell structure for three parallel forces. It may be noted that the moments at A, B are of opposite sign to those that would be developed at the ends of a built-in beam by a downward central load. The optimum layout for the case of reversed moments at A and B remains to be determined. In the particular case when there is no moment across the central section, that is the case $M = Pa/2$, the optimum layout is as shown in Figure 7. In the associated virtual deformation, the space outside the circular regions does not move while inside the circular regions the displacement is purely circumferential.

Acknowledgment. The author would like to thank D. E. Carlson for helpful discussions. The manuscript was typed by Mrs. R. A. Mathine.

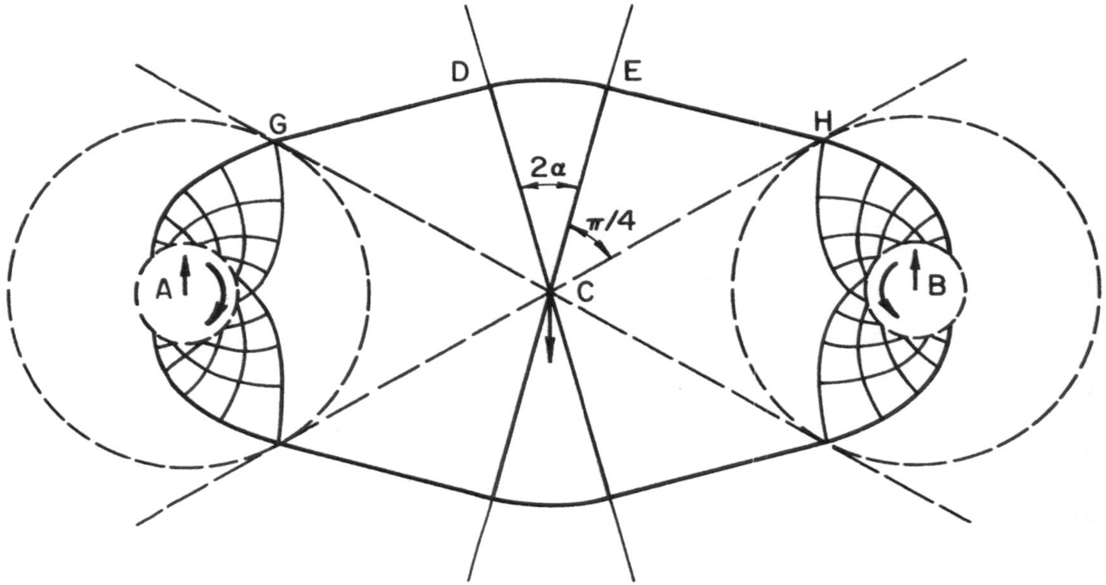

Figure 6. Bending with central load

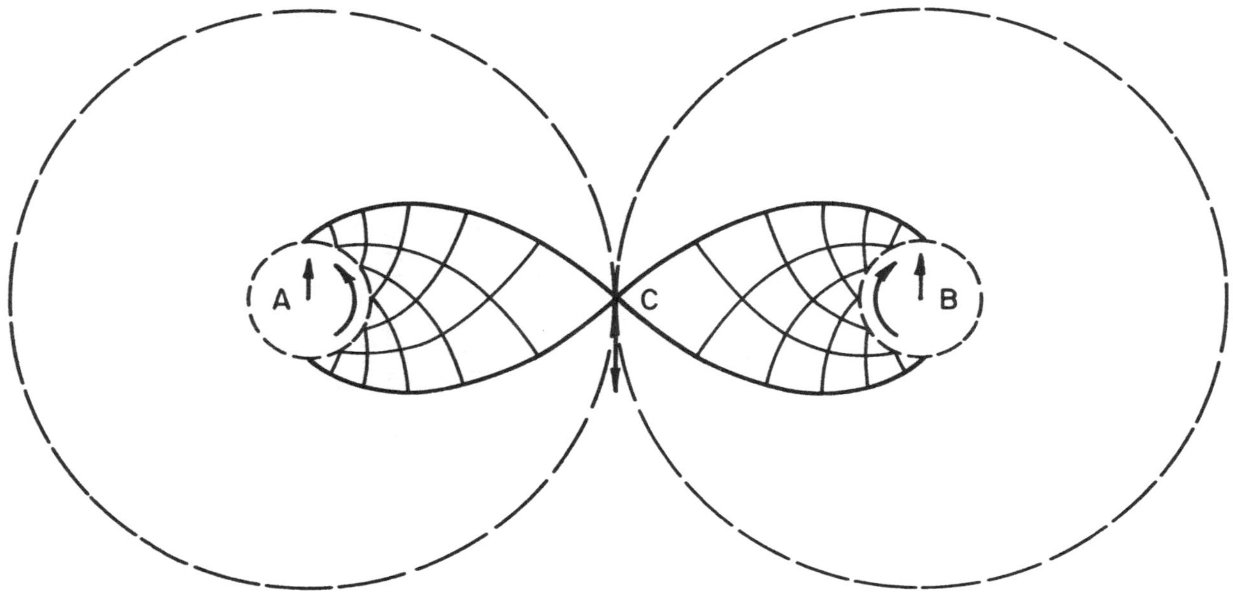

Figure 7. Bending with central load and zero central moment

REFERENCES

1. Michell, A. G. M. The Limits of Economy in Frame-Structures. Phil. Mag. 8, 589-597 (1904).

2. Drucker, D. C., and Shield, R. T. Design for Minimum Weight. Proc. 9th International Congress of Applied Mechanics, Brussels 1956, pp. 212-222.

3. Drucker, D. C., and Shield, R. T. Bounds on Minimum Weight Design. Q. Appl. Math. 15, 269-281 (1957).

4. Shield, R. T. On the Optimum Design of Shells. J. Appl. Mech. 27, 316-322 (1960).

5. Prager, W., and Taylor, J. E. Problems of Optimal Structural Design. J. Appl. Mech. 35, 102-106 (1968).

6. Shield, R. T., and Prager, W. Optimal Structural Design for Given Deflection. J. Appl. Math. Phys. (ZAMP) 21, 513-523 (1970).

7. Shield, R. T. Optimum Design Methods for Structures. Plasticity, Proc. 2nd Symp. Naval Struct. Mechanics, Providence 1960, pp. 580-591.

8. Wasiutyński, Z., and Brandt, A. The Present State of Knowledge in the Field of Optimum Design of Structures. Appl. Mech. Rev. 16, 341-350 (1963).

9. Sheu, C. Y., and Prager, W. Recent Developments in Optimal Structural Design. Appl. Mech. Rev. 21, 985-992 (1968).

10. Prager, W. Optimization of Structural Design. J. Optimization Theory and Applic. 6, 1-21 (1970).

11. Anderson, C. A., and Shield, R. T. On the Validity of the Plastic Theory of Structures for Collapse under Highly Localized Loading. J. Appl. Mech. 33, 629-636 (1966).

12. Drucker, D. C., Prager, W., and Greenberg, H. J. Extended Limit Design Theorems for Continuous Media. Q. Appl. Math. 9, 381-389 (1952).

13. Prager, W. General Theory of Limit Design. Proc. 8th International Congress of Applied Mechanics, Istanbul 1952.

14. Mróz, Z. The Load Carrying Capacity and Minimum Weight Design of Annular Plates. Rozpr. Inżyn. (Engin. Trans., Warsaw) 114, 605-625 (1958).

15. Onat, E. T., Schumann, W., and Shield, R. T. Design of Circular Plates for Minimum Weight. J. Appl. Math. Phys. (ZAMP) 8, 485-499 (1957).

16. Prager, W., and Shield, R. T. Minimum Weight Design of Circular Plates under Arbitrary Loading. J. Appl. Math. Phys. (ZAMP) 10, 421-426 (1959).

17. Shield, R. T. Plate Design for Minimum Weight. Q. Appl. Math. 18, 131-144 (1960).

18. Shield, R. T. Optimum Design Methods for Multiple Loading. J. Appl. Math. Phys. (ZAMP) 14, 38-45 (1963).

19. Save, M. A., and Shield, R. T. Minimum-Weight Design of Sandwich Shells Subjected to Fixed and Moving Loads. Proc. 11th International Congress of Applied Mechanics, Munich 1964, pp. 341-349.

20. Save, M. A. Some Aspects of Minimum-Weight Design. Engineering Plasticity, Cambridge Univ. Press 1968, pp. 611-626.

21. Prager, W., and Shield, R. T. Optimal Design of Multi-Purpose Structures. Int. J. Solids Structures 4, 469-475 (1968).

22. Hemp, W. S. Studies in the Theory of Michell Structures. Proc. 11th International Congress of Applied Mechanics, Munich 1964, pp. 621-628.

23. Hegemeir, G. A., and Prager, W. On Michell Trusses. Int. J. Mech. Sci. 11, 209-215 (1969).

24. Prager, W. On a Problem of Optimal Design. Non-homogeneity in Elasticity and Plasticity, Pergamon Press 1959, pp. 125-132.

25. Hu, T. C., and Shield, R. T. Minimum-Volume Design of Discs. J. Appl. Math. Phys. (ZAMP) 12, 414-433 (1961).

26. Cox, H. L. The Design of Structures of Least Weight, Pergamon Press 1965.

27. Cox, H. L. The Theory of Design. Aeronautical Research Council Report 19791, Great Britain (1958).

OPTIMIZATION PROBLEMS IN HYDROFOIL PROPULSION*

Th. Yao-tsu Wu, Allen T. Chwang
California Institute of Technology, Pasadena, California
and Paul K. C. Wang
University of California at Los Angeles

This paper attempts to apply the principle of control theory to investigate the possibility of extracting flow energy from a fluid medium by a flexible hydrofoil moving through a gravity wave in water, or by an airfoil in gust. The present optimization consideration has led to the finding that although the flexible hydrofoil may have an infinite number of degrees of freedom, the optimum shape problem is nevertheless a finite-dimensional one. The optimum shape sought here is the one which minimizes the required power subject to the constraint of fixed thrust. A primary step towards the solution is to reduce the problem to one of minimizing a finite quadratic form; after this reduction the solution is determined by the method of variational calculation of parameters. It is found that energy extraction is impossible if the incident flow is uniform, and may be possible when the primary flow contains a wave component having a longitudinal distribution of the velocity component normal to both the mean direction of flight and the wing span. When such waves of sufficiently large amplitude are present, not only flow energy but also a net mechanical power can be extracted from the surrounding flow.

* This paper includes further extension to that which was originally presented at the Symposium.

1. Introduction

Some previous observations on fish swimming and bird flight seem to suggest that some species may have learned, through experience, to acquire the key to high performance by executing the optimum movement that may be of great interest to control theory related to fluid mechanics. An especially intriguing aspect of the optimization problem concerns with the possibility of extracting energy from surrounding flow by an oscillating lifting surface (such as the fish body and fins, bird wings, and artificial wings like airfoil and hydrofoil) and its associated effect on the control of motion.

This general problem has been explored to various degrees of generality. Based on the approximation of potential flow with small amplitude, it has been found by Lighthill (1960) for slender bodies, and by Wu (1961) for two-dimensional plates, that if the basic flow is uniform, energy is always imparted by an oscillating wing to the surrounding fluid, and an extraneous mechanical work must therefore be continuously supplied to maintain the motion. Even though it is impossible in this case to extract energy from the flow field, the highest possible hydromechanical efficiency that can be attained by a wing, subject to delivering a given forward thrust, can be very high, as found by Wu (1971 b, c) for the two-dimensional plate and a slender lifting surface.

As was subsequently pointed out by Wu (1972), the situation becomes drastically different when the basic flow is no longer uniform, but contains a wave component, such as gravity waves in water, or wavy gust in air. The contention that the wave energy stored in a fluid medium can be utilized to assist propulsion has been suggested by intuitive observations. Sea gulls and pelicans have been observed to skim ocean waves over a long distance without making noticeable flapping motions (save some gentle twisting) of their wings. In an extensive study of the migrating salmon, Osborne (1960) found that the increased flow rate in a swollen river did not slow the salmon down (for known biochemical energy expended during the travel) by that much a margin as would be predicted by the law of resistance in proportion to the square of their velocity relative to the flowing water. Several possible explanations were conjectured by Osborne, including the prospect that the flow energy associated with the eddies in river could be converted to generate thrust. To explore this possibility Wu (1972) introduced an energy consideration to an earlier study of Weinblum (1954) on the problem of heaving and pitching of a rigid hydrofoil in regular water waves. It was found that the greatest possible rate of energy extraction is provided by the optimum mode of heaving and pitching. When waves of sufficiently large amplitude are present, not only flow energy but also a net mechanical power can be extracted from the wave field.

In the present study this problem is further generalized by allowing the hydrofoil to be flexible so as to admit an infinitely many degrees of freedom (of small amplitudes). This general problem merits study for several reasons. First, it is of a theoretical interest to find out how much improvement in the hydromechanical efficiency and energy extraction can be gained by admitting the additional degrees of freedom. Second, the results of the present study of energy transfer between an oscillating body and surrounding stream can be useful to the development of control theory for hydrofoil ships and to the analysis of flutter phenomena. In the case of flutter in a uniform stream, it is usually assumed that the engine maintains the constant forward speed regardless of the flutter-created inertial drag. In a wavy stream, however, the flutter may create a propulsive thrust, which may amplify further instability and a self-excited flutter may develop. Some of these aspects have already been observed by Küssner (1935) and Garrick (1936, 1957); this paper is aimed at the general case of propulsive energy balance.

Further, from the standpoint of development of control theory, the present problem also merits study in its own right since it presents some new features and difficulties that apparently do not confirm with the known classical cases. A brief description can be given as follows. Section 2 presents the general (linearized) theory for a two-dimensional hydrofoil oscillating in waves, which is applied in Section 3 to the general case of a flexible plate wing. In Section 4 the problem of optimum motion is formulated as to find a hydrofoil profile that minimizes the energy loss C_E subject to a constrained thrust coefficient C_T. It is shown that although the flexible hydrofoil may have infinitely many degrees of freedom, C_E and C_T can be reduced to quadratic forms of finite dimensions. After this crucial step the optimization problem reduces to one defined on a three-dimensional vector space ($\zeta_1, \zeta_2, \zeta_3$). With this drastic reduction it is possible to show that an optimal solution does not exist unless appropriate bounds are imposed on the independent variables ζ_n's. Under this condition the optimal solution is determined and compared with the previous special cases. It is felt that the present method of solution is still heuristic, to some extent, for much of the intuitive physical picture was relied on for guidance. It is with the hope to stimulate further development of the general theory for this class of control problem that the present study is presented before this Symposium.

2. Two-dimensional Hydrofoil Oscillating in Waves

With specific applications in view we consider the basic flow to be a sinusoidal gravity wave of small amplitude in water of finite depth, H, in which a two-dimensional hydrofoil of chord 2ℓ moves horizontally with velocity U while submerged at a mean depth h_1 underneath the free surface. In terms of the body coordinate system (x, y), the wave profile of the basic flow (see Fig. 1) may be written as

$$y = h_1 + \mathrm{Re}\,[\, a\, e^{i(\omega_0 t - kx)}\,] \quad , \tag{1}$$

the wave amplitude, a, being assumed small such that $ka \ll 1$. The corresponding velocity $(U + u_0,\ v_0)$ of the wave field, by classical theory, is

$$u_0 - iv_0 = A_* \cos[\, k(x + iy + ih_2) - \omega_0 t\,] \ , \tag{2}$$

where $h_2 = H - h_1$ is the distance from the bottom and ω_0 is the encounter frequency

$$\omega_0 = \omega_* \pm kU \quad , \tag{3}$$

$$\omega_*^2 = gk \tanh kH \quad , \qquad A_*/a = (2gk/\sinh 2kH)^{\frac{1}{2}} \quad , \tag{4}$$

where g is the gravitational acceleration and in (3), the + sign is for heading sea, and - sign for following the waves. In particular, the y-component wave velocity at the x-axis (which coincides with the mean position of the hydrofoil), denoted by $v_0(x, 0, t) \equiv V_0(x, t)$, is

$$V_0(x, t) = iA_0\, e^{i(\omega_0 t - kx)} \quad , \qquad A_0 = A_* \sinh kh_2 \ . \tag{5}$$

Here and henceforth, the real part of a complex expression will be understood for physical interpretation.

Since the problem of central interest at hand is to determine the effect of a waving stream on the propulsive performance of a hydrofoil in unsteady motion, we shall further assume, for simplicity, that the hydrofoil is located sufficiently far from both the free water surface and solid bottom so as to curtail the complicated (but only secondary) corrections due to these boundary effects. This condition would be nearly satisfied if the hydrofoil is at a distance more than two chords away from each of these boundaries, that is for the chord $2\ell < \frac{1}{2}\max\{h_1, h_2\}$, this estimate being inferred by the known results of the steady flow case (see Wu, 1954) which is assumed to remain valid in the unsteady case. As an additional simplifying assumption, the ratio $\epsilon = A_0/U$ of the magnitude of the orbital wave velocity to the mean free stream velocity is taken to be small so that the x-component

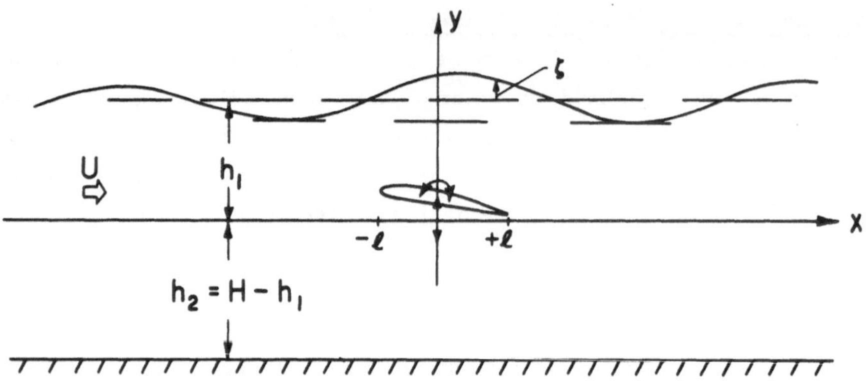

FIGURE 1

orbital velocity, u_o, may be neglected in comparison with U in formulating the present linear theory. Although the simple water wave is chosen as a concrete example, it makes little difference to the subsequent discussion if other kinds of wavy streams are considered as long as the transverse velocity of the basic flow can be represented by equation (5). For arbitrary V(x, t) the result can be obtained by the Fourier synthesis of this fundamental case.

The hydrofoil (or airfoil) is assumed to be thin, though sufficiently rounded at the leading edge to keep the flow from being separated there. The foil-thickness effect is then only secondary and will be further disregarded in this study. For brevity, the semi-chord, ℓ, of the hydrofoil will be normalized to unity as the reference length. The unsteady motion of the hydrofoil assumes the fundamental form

$$y = h(x,t) = \hat{h}(x)\, e^{i\omega t} \qquad (-1 < x < 1) , \tag{6}$$

where the circular frequency ω is arbitrary, and $\hat{h}$ may be a complex function of x (with respect to $i = \sqrt{-1}$ in the time factor). With the resultant flow velocity denoted by $(U + u_o + u_1,\ v_o + v_1)$, the linearized boundary condition that the flow be always tangential to the moving body surface requires

$$v_1^{\pm}(x,t) \equiv V_1(x,t) = V(x,t) - V_o(x,t) \qquad (|x| < 1) \tag{7a}$$

$$V(x,t) = D\,h(x,t) , \qquad D \equiv \partial/\partial t + U\,\partial/\partial x , \tag{7b}$$

where $V_o(x,t)$ is given by equation (5), and $v_1^{\pm}(x,t)$ signifies $v_1(x, 0\pm, t)$. Like in the uniform stream case, it is convenient to use the acceleration potential, defined by

$$\phi = (p_\infty - p)/\rho = \phi_0 + \phi_1 , \tag{8}$$

as a new dependent variable, particularly since it is continuous throughout the fluid. It is related to the velocity by

$$Du_o = \partial\phi_o/\partial x , \quad Dv_o = \partial\phi_o/\partial y - g ; \quad Du_1 = \partial\phi_1/\partial x , \quad Dv_1 = \partial\phi_1/\partial y, \tag{9}$$

on linearized theory. The component ϕ_o, which gives the pressure distribution in the primary wave field, can be readily obtained by integration of the first two equations in (9); it gives no hydrodynamic force (except a bouyancy) or moment on the hydrofoil since it is continuous across the plate. The effect of waving stream on the hydrodynamic performance comes with calculations on ϕ_1, explicitly through the term $V_o(x,t)$ in condition (7). The problem of ϕ_1 is specified by the boundary conditions

$$\partial\phi_1^{\pm}/\partial y = D\, V_1(x,t) \qquad (|x| < 1,\ y = 0\pm), \tag{10a}$$

$$\phi_1^{\pm} = 0 \qquad (|x| > 1,\ y = 0\pm), \tag{10b}$$

together with the Kutta condition that $\phi_1^{\pm} = 0$ at $x = 1$, and that ϕ_1 vanishes at infinity. Condition (10a) follows from substituting (7a) into the last equation of (9); and (10b) is a consequence of the pressure being continuous in the flow and the fact that ϕ_1 is odd in y.

The solution to this mixed-type boundary problem of ϕ_1 is known (see Wu, 1971a); in particular, the value of $\phi_1(x, 0\pm, t)$ at the plate is given by

$$\phi_1^{\pm}(x,t) = \pm\frac{U}{2}a_0\left(\frac{1-x}{1+x}\right)^{\frac{1}{2}} \pm \frac{1}{\pi}\int_{-1}^{1}\left(\frac{1-x^2}{1-\xi^2}\right)^{\frac{1}{2}}\frac{\psi_1(\xi,t)}{\xi - x}\,d\xi \tag{11}$$

$$\psi_1(x,t) = -D\int_{-1}^{x} V_1(\xi,t)\,d\xi \,, \tag{12}$$

$$a_0 = [\, b_1 - (b_0 + b_1)\Theta(\sigma)] - [\, b_1' - (b_0' + b_1')\Theta(\sigma_0)] \,, \tag{13}$$

$$b_n = \frac{2}{\pi}\int_0^{\pi} V(\cos\theta, t)\cos n\theta\, d\theta \qquad (x = \cos\theta,\ \ n = 0, 1, 2, \dots)\,, \tag{14}$$

$$b_n' = \frac{2}{\pi}\int_0^{\pi} V_0(\cos\theta, t)\cos n\theta\, d\theta = 2A_0(-i)^{n-1} J_n(k)\, e^{i\omega_0 t} \,, \tag{15}$$

$$\Theta(\sigma) = \mathfrak{F}(\sigma) + i\,\mathfrak{G}(\sigma)\,, \qquad \sigma \equiv \omega\ell/U\,, \qquad \sigma_0 \equiv \omega_0\ell/U\,, \tag{16}$$

$$\sigma_0 = (\delta\kappa)^{\frac{1}{2}} \pm \kappa\,, \qquad \delta \equiv (g\ell/U^2)\tanh kH\,, \qquad \kappa \equiv k\ell\,. \tag{17}$$

Here the integral in equation (11) assumes its Cauchy principal value; $J_n(k)$ is the Bessel function of the first kind; $\Theta(\sigma)$ is the Theodorsen function, $\mathfrak{F}$ and $\mathfrak{G}$ being its real and imaginary parts(for a tabulation of Θ, see Luke and Dengler, 1951); σ is the reduced frequency of the body motion, σ_0 the wave reduced frequency, both being based on the half-chord ℓ. The function $\sigma_0(\kappa)$ in equation (17) is the non-dimensional form of equation (3). We shall write κ as k since $\ell = 1$.

The differential lift distribution along the chord is clearly

$$\mathcal{L}(x,t) = p^-(x,t) - p^+(x,t) = 2\rho\phi_1^{+}(x,t) \qquad (|x| < 1)\,. \tag{18}$$

The integral representation of the lift L and the moment M (about the mid-chord, positive in the nose-up sense) are

$$L = \int_{-1}^{1}\mathcal{L}(x,t)\,dx \,, \tag{19}$$

$$M = -\int_{-1}^{1} \mathcal{L}(x,t)\, x\, dx \ . \tag{20}$$

Also, the formulas for calculating the thrust, T, the power required for maintaining the motion, P, and the kinetic energy imparted to the fluid, E, remain the same as in the uniform stream case (Wu, 1971a),

$$T = \int_{-1}^{1} \mathcal{L}(x,t)\, h_x\, dx + S \ , \quad S = \tfrac{1}{2}\pi\rho(\mathrm{Re}\ a_0)^2 \ , \tag{21}$$

$$P = -\int_{-1}^{1} \mathcal{L}(x,t)\, h_t\, dx \ , \tag{22}$$

$$E = -\int_{-1}^{1} \mathcal{L}(x,t)(h_t + Uh_x)\, dx - SU \ , \tag{23}$$

where the subscripts designate partial differentiation, S represents the leading-edge suction, which now includes the contribution from the wave component. From the above expressions we note that the energy balance can be expressed as

$$P = TU + E \ , \tag{24}$$

which is formally the same as in the uniform stream case. However, unlike the case of uniform free stream, the time average of E here is no longer always positive, and we shall see that energy can be extracted from the waves when E becomes negative.

3. Flexible Plate Wing

We shall consider the general case when the motion is periodic in time, as prescribed by (6), with arbitrary amplitude function $\hat{h}(\cdot)$, such as any one that can be performed by a completely flexible plate. Substituting the differential lift $\mathcal{L}(x,t)$ given by (11) - (18) in (19) and (20), we obtain the total lift L, and the moment M, as

$$L = \pi\rho U\,\{a_0 - (b_1 - b_1') - \tfrac{1}{2}i\sigma(b_0 - b_2) + \tfrac{1}{2}i\sigma_0(b_0' - b_2')\} \ , \tag{25}$$

$$M = \tfrac{1}{2}\pi\rho U\,\{a_0 + (b_2 - b_2') + \tfrac{1}{4}i\sigma(b_1 - b_3) - \tfrac{1}{4}i\sigma_0(b_1' - b_3')\} \ . \tag{26}$$

The corresponding results for the thrust T, power P, and energy loss E can be obtained by substituting (11) - (18) in (21) - (23) and by following the same procedure as that used for the uniform stream case by Wu (1971a, in deriving his equations 44 - 46); the intermediate manipulation again can be considerably simplified by making use of the relationship

$$\int_{-1}^{1} f'(x)\,dx \oint_{-1}^{1}\left(\frac{1-x^2}{1-\xi^2}\right)^{\frac{1}{2}} \frac{g(\xi)\,d\xi}{\xi - x} = \int_{-1}^{1} g'(x)\,dx \oint_{-1}^{1}\left(\frac{1-x^2}{1-\xi^2}\right)^{\frac{1}{2}} \frac{f(\xi)\,d\xi}{\xi - x} , \tag{27}$$

where $f(x)$, $g(x)$ are two arbitrary functions, provided they and their derivatives $f'(x)$, $g'(x)$ are continuous in $-1 \leqslant x \leqslant 1$. The mean values of thrust $\overline{T}$, power $\overline{P}$, and energy loss $\overline{E}$ can be deduced by averaging T, P, and E over a long time period. Two different cases arise according as $\omega = \omega_0$ or $\omega \neq \omega_0$.

(i) When $\omega = \omega_0$, that is when the wing oscillates at the wave encounter frequency, the two motions are correlated. In this case we obtain $\overline{T}$ as

$$\overline{T} = \frac{\pi}{4}\rho\, \mathrm{Re}\{(a_0 + b_0 - \dot{\beta}_0)(a_0^* - b_1^* + \dot{\beta}_1^*) - b_0'\dot{\beta}_1^* - b_1'\dot{\beta}_0^* + \dot{\beta}_0\dot{\beta}_1^* + 2I\} , \tag{28a}$$

$$I = \frac{2}{\pi^2}\int_{-1}^{1}\oint_{-1}^{1}\left(\frac{1-x^2}{1-\xi^2}\right)^{\frac{1}{2}} \frac{V(x,t)\,V_o^*(\xi,t)}{\xi - x}\,dx\,d\xi , \tag{28b}$$

where the superscript * denotes the complex conjugate, $\dot{\beta}_n \equiv d\beta_n(t)/dt$, and

$$\beta_n(t) = \frac{2}{\pi}\int_0^{\pi} h(x,t)\cos n\theta\, d\theta \qquad (x = \cos\theta,\ \ n = 0, 1, 2, \dots) , \tag{28c}$$

hence $\dot{\beta}_n = i\omega\beta_n$ when h is given by (6). The mean power $\overline{P}$ can be shown, after some manipulation, to have the following expression

$$\overline{P} = \frac{\pi}{4}\rho U\, \mathrm{Re}\{(a_0 + b_0 - b_0')\dot{\beta}_1^* + (b_1 - b_1' - a_0)\dot{\beta}_0^* + 2I_1\} , \tag{29a}$$

$$I_1 = -\frac{2}{U\pi^2}\int_{-1}^{1}\oint_{-1}^{1}\left(\frac{1-x^2}{1-\xi^2}\right)^{\frac{1}{2}} \frac{\partial V_o^*(x,t)/\partial t}{\xi - x}\left(\int_{-1}^{\xi} V(\eta,t)\,d\eta\right)dx\,d\xi . \tag{29b}$$

By substituting in (29b) the relationship

$$\partial V_0(x,t)/\partial t = -(\omega/k)\,\partial V_0/\partial x ,$$

(see (5), with $\omega = \omega_0$ for the present case), and applying the formula (27), it immediately follows that

$$I_1 = (\sigma/k)\, I , \tag{29c}$$

where I is given by (28b). Whence, by (24), $\overline{E} = \overline{P} - U\overline{T}$, or

$$\overline{E} = \frac{\pi}{4}\rho U\, \mathrm{Re}\{(a_0 + b_0)(b_1^* - a_0^*) + 2(\sigma/k - 1)I\} . \tag{30}$$

We note that, upon substituting (13) in (30), the first term on the right-hand side of (30) involves only the first two Fourier coefficients of V, in the particular combination of $(b_0 + b_1)$. However, the second term in (30) with I, which results from the interaction between the wave action and body motion, involves all the Fourier coefficients of V since the integral I has the following Fourier-Bessel expansion

$$I = A_o e^{-i\omega t} \sum_{n=1}^{\infty} (i)^{n+1} J_n(k) (b_{n+1} - b_{n-1}) . \tag{31}$$

We further note that the expressions for $\overline{T}$ and $\overline{P}$ involve, in addition to the b_n's, also the first two Fourier coefficients, β_o and β_1, of h(x, t).

To facilitate the subsequent consideration of the optimum shape problem, it is useful to recast the above expressions for $\overline{T}$, $\overline{P}$, and $\overline{E}$ in terms of certain inner products. Let $\mathfrak{H}$ denote a subset of the complex Hilbert space $L_2[-1,1]$

$$\mathfrak{H} \equiv \left\{ f \in L_2[-1,1] : \frac{2}{\pi} \int_{-1}^{1} |f(x)|^2 (1-x^2)^{-\frac{1}{2}} dx < \infty \right\} \tag{32a}$$

and let the inner product between f(•) and g(•) on $\mathfrak{H}$ be defined by

$$<f,g> \equiv \frac{2}{\pi} \int_{-1}^{1} f(x) g^*(x) (1-x^2)^{-\frac{1}{2}} dx = <g,f>^* , \quad (f, g \in \mathfrak{H}) \tag{32b}$$

where the weighting function $(1-x^2)^{-\frac{1}{2}}$ is introduced in order to convert the Fourier coefficients into the inner product form. Any two functions f, g in $\mathfrak{H}$ will be said to be orthogonal on $\mathfrak{H}$ if $<f, g> = 0$.

Substituting (13) - (16) and (28c) in (29) - (30), we obtain the mean coefficients of thrust, power, and energy loss, defined by $(C_P, C_E, C_T) = (\overline{P}, \overline{E}, \overline{T}U)/(\frac{1}{4}\pi\rho U^3 \ell)$, in terms of the inner products as

$$C_P = \mathrm{Re}\{ -i\sigma[<v,f_1> - 2\epsilon (J_1 + iJ_o)] <g_1, \hat{h}> + 2\epsilon (\sigma/k) <v, g_2> \} , \tag{33}$$

$$C_E = \mathrm{Re}\{ B(\sigma)|<v,f_1>|^2 + 2\epsilon (1-2\Theta)(W_1 + iW_2) <v,f_1> + 2\epsilon (\sigma/k - 1) <v, g_2> - 4\epsilon^2 W^2 \}, \tag{34}$$

where

$$v(x) = e^{-i\omega t} V(x,t)/U , \qquad \hat{h}(x) = e^{-i\omega t} h(x,t) , \tag{35a}$$

$$f_1(x) = 1 + x , \qquad g_1(x) = (1-\Theta)x + \Theta , \qquad \epsilon = A_o/U , \tag{35b}$$

$$\Theta(\sigma) = \mathfrak{F}(\sigma) + i\,\mathfrak{G}(\sigma) , \qquad B(\sigma) = \mathfrak{F} - (\mathfrak{F}^2 + \mathfrak{G}^2) , \tag{35c}$$

$$W_1 - iW_2 = J_1(k)[1 - \Theta(\sigma)] - iJ_0(k)\Theta(\sigma) , \qquad W^2 = W_1^2 + W_2^2 , \tag{35d}$$

$$g_2(x) = \frac{i}{\pi}(1-x^2)\int_{-1}^{1}\frac{e^{-ik\xi}\,d\xi}{(1-\xi^2)^{\frac{1}{2}}(\xi-x)} \quad . \tag{35e}$$

In the above, as well as in the sequel, the argument k of the Bessel functions $J_n(k)$ will be understood unless otherwise designated. The mean thrust coefficient is simply (the coefficient form of (24))

$$C_T = C_P - C_E \quad . \tag{36}$$

Another flow quantity of interest is the mean leading-edge-suction coefficient, $C_S = \overline{S}/\frac{1}{4}\pi\rho U^2 \ell$. From (21), (13), (14) we obtain

$$C_S = |\langle v, f_1 \rangle \Theta - \langle v, f_o \rangle + 2\epsilon\,(W_1 - iW_2)|^2 \quad , \tag{37a}$$

where $\quad f_o(x) = x \qquad (-1 \leqslant x \leqslant 1)$.

As suggested by Lighthill (1969, 1970), the ratio C_S/C_T provides a measure of the relative strength of the leading-edge suction; moderate and large values of C_S/C_T (as compared to unity) suggest a tendency that the flow would separate, or stall, near the leading edge (such a category of separated flow would be quite different from the completely wetted flow as assumed here).

(ii) $\omega \neq \omega_o$ --- In this case the mean product of $\exp(i\omega t)$ and $\exp(\pm i\omega_o t)$ vanishes as the body motion and wave action become uncorrelated. Consequently the terms which are linear in ϵ in (33) and (34) drop out of the expressions for C_P and C_E; further, W^2 in (34) then assumes its value at σ_o. The corresponding C_S likewise becomes

$$C_S = |\langle v, f_1 \rangle \Theta - \langle v, f_o \rangle|^2 + 4\epsilon^2 W^2(\sigma_o, k) \quad . \tag{38}$$

The result of this case therefore reduces virtually to the case of uniform stream except for the additional term $(-4\epsilon^2 W^2)$ in the expression for C_E and $(4\epsilon^2 W^2)$ in C_S. These added terms indicate that energy is invariably being supplied by the primary wave, through the mechanism of generating a greater leading-edge suction, at no expense of C_P. It thus follows that for C_P fixed, C_T becomes greater and C_E smaller (hence higher efficiency) with increasing wave action (greater ϵW). The energy gain in this case, however, is always accompanied by an appreciable increase in the leading-edge suction, suggesting an easier leading-edge stall. When the suction is required to remain reasonably small, the optimum motion and the corresponding improvement of efficiency are not significantly different from the uniform stream case which has been discussed earlier by Wu

(1971b). For this reason this second case will not be further pursued here.

4. The Optimum Motion ($\omega = \omega_0$)

The present problem of optimum motion is formulated especially to analyze the interaction between the body motion and wave action; it can be stated as follows:
Given a reduced frequency $\sigma > 0$ (hence also the wave number k, see (3)) and a thrust coefficient $C_{T,0} > 0$, find a velocity profile v, or a hydrofoil profile $\hat{h}$ in the set $\mathcal{H}$ (defined by (32a)) such that C_E is minimized subject to the constraint

$$C_T = C_{T,0} > 0 , \tag{39}$$

assuming that the wing oscillates at the wave encounter frequency.

It is desirable to choose C_T (rather than C_P or C_E) to be a constrained quantity since a constant thrust is required to overcome the (nearly constant) viscous drag if the uniform forward motion is to be maintained. No additional constraints are imposed here on the total lift L and moment M for balancing the rectilinear and angular recoils of the flexible plate (see Wu, 1971a, Eqs. (56a,b)); this choice is made for two reasons. First, when a body structure consists of components other than the flexible plate, the recoil consideration must take the motion of the entire body into account. Second, even when the wing alone comprises a self-propelling body in its entirety, there will still be other degrees of freedom left to be used to satisfy the recoil conditions, if desired, as we shall see later.

In choosing the independent functionals for the optimization calculation, we note that only two of C_P, C_E, C_T are independent since they are related by (36). There are great advantages in the choice of C_P and C_E as the independent functionals of v and $\hat{h}$ because C_E, in particular, does not involve $\hat{h}$, and C_P is also simpler in expression than C_T. In the expression (34) for C_E, the first term on the right-hand side is the same as in the uniform-stream case (see Wu, 1971b, Eq. (13)); it is always non-negative since $B(\sigma) > 0$ for $\sigma > 0$. The second and third terms, which are bilinear in ϵ and v, represent the body-wave interaction. The last term, which is proportional to ϵ^2, is solely due to the wave action. This result actually proves the statement that extraction of energy from the surrounding flow by an oscillating flexible wing is impossible if the incident flow is uniform. In the presence of a primary wave, with appropriate v and increasing wave parameter ϵ, the last three terms in (34) may become negative and numerically so large as to reduce C_E at first, and C_P eventually, to negative values, as will be seen later. The case of $C_P < 0$ signifies the operation in which a mechanical power is received by the body, instead of being consumed by it, through a favorable extraction of the wave energy. In spite of these possibilities, we shall still continue to use the Froude efficiency

$$\eta = C_T/C_P = C_{T,0}/C_P = (1 + C_E/C_{T,0})^{-1} \tag{40a}$$

as a measure of the hydromechanical performance. Aside from its usual significance for $0 < \eta < 1$, now we may have new generalized interpretations as follows:

$$\text{(i)}\ \eta > 1 \quad \text{for} \quad C_E < 0,\ C_P > 0; \tag{40b}$$

$$\text{(ii)}\ \eta < 0 \quad \text{for} \quad C_E < C_P < 0. \tag{40c}$$

Another step of primary importance is to choose the independent function for the optimization calculation. Although either v or h may serve as an independent function (since they are related by a differential equation (7b)), the advantage of taking v is clear, as was noted by Wu (1971b, section 6) in discussing the optimum shape of a flexible plate oscillating in a uniform stream. As another reason, we note that in the present formulation, an inner product of h with a given $f(\cdot)$ can be converted into an equivalent one involving v, whereas the converse is generally impossible.

Accepting v as the independent function, we proceed to recast the inner product $< g_1, \hat{h} >$ in (33) in terms of v. By (35a) and (7b), $\hat{h}$ and v are related by

$$(d/dx + i\sigma)\, \hat{h}(x) = v(x) \qquad (|x| < 1)\ , \tag{41a}$$

which has the general integral as

$$\hat{h}(x) = \int_{-1}^{x} e^{-i\sigma(x-\xi)}\, v(\xi)\, d\xi + \hat{h}_{-1} e^{-i\sigma(x+1)}\ , \tag{41b}$$

where $\hat{h}_{-1}$ is an arbitrary integration constant. Substituting (41b) and (35b) in $< g_1, \hat{h} >$, and integrating by parts, we obtain

$$< g_1, \hat{h} > = < g_3, v > + C_1 - iC_2\ , \tag{42a}$$

where

$$g_3(x) = (1 - x^2)^{\frac{1}{2}} \int_{x}^{1} e^{-i\sigma(x-\xi)} (1 - \xi^2)^{-\frac{1}{2}} g_1(\xi)\, d\xi\ , \tag{42b}$$

$$C_1 - iC_2 = 2i\hat{h}^*_{-1}\, e^{i\sigma} [\, J_1(\sigma)(1 - \Theta) - i\Theta J_0(\sigma)]\ . \tag{42c}$$

Consequently (33) becomes

$$C_P = \text{Re}\, \{ -i\sigma[< v, f_1 > - 2\epsilon\, (J_1 + iJ_0)] [< g_3, v > + C_1 - iC_2] + 2\epsilon\ (\sigma/k) < v, g_2 > \}\ . \tag{33$'$}$$

Now the expression for C_P in $(33)'$ and C_E in (34) are both expressed in terms of v and contain only three inner products: $<v,f_1>$, $<v,g_2>$, and $<g_3,v>$.

Since f_1, g_2, g_3 are not mutually orthogonal on $\mathcal{H}$, we next construct a set of three orthogonal functions, f_1, f_2, f_3 say(there being no need here to normalize them), by the Schmidt scheme:

$$f_1 = 1 + x \qquad [\, <f_1,f_1> = 3\,] \tag{43a}$$

$$g_2 = a_1 f_1 + f_2 \,, \tag{43b}$$

$$g_3 = a_2 f_1 + a_3 f_2 + f_3 \,, \tag{43c}$$

such that

$$<f_i,f_k> = 0 \qquad (i \neq k)\,. \tag{44}$$

The coefficients a_n are determined by the orthogonality condition (44) as

$$a_1 = <g_2,f_1>/<f_1,f_1> = \tfrac{1}{3}<g_2,f_1> = \tfrac{1}{3}[\,2J_1(k) - iJ_2(k)]\,, \tag{45a}$$

$$a_2 = <g_3,f_1>/<f_1,f_1> = \frac{2}{3\sigma^2}\{\Theta[\,1+i\sigma - e^{i\sigma}J_0(\sigma)] + i(1-\Theta)[\tfrac{\sigma}{2} - e^{i\sigma}J_1(\sigma)]\}, \tag{45b}$$

$$a_3<f_2,f_2> = <g_3,f_2> = <g_3,g_2> - a_1^* <g_3,f_1> = <g_3,g_2> - 3a_1^* a_2\,. \tag{45c}$$

By separate calculations,

$$<f_2,f_2> = <g_2,g_2> - a_1<f_1,g_2> - a_1^*<g_2,f_1> + a_1 a_1^* <f_1,f_1> = <g_2,g_2> - 3a_1 a_1^*\,,$$

$$<g_2,g_2> = \frac{2}{\pi^3}\int_{-1}^{1}(1-x^2)^{3/2}dx\int_{-1}^{1}\frac{e^{-ik\xi}\,d\xi}{(1-\xi^2)^{\frac{1}{2}}(\xi-x)}\int_{-1}^{1}\frac{e^{ik\eta}\,d\eta}{(1-\eta^2)^{\frac{1}{2}}(\eta-x)}$$

$$= 1 - J_0^2(k) + 2J_1^2(k) - 2J_0(k)J_2(k)\,,$$

which can be shown by successive interchange of the order of integration and by making use of the Poincaré-Bertrand formula, and hence

$$<f_2,f_2> = 1 - J_0^2(k) + \tfrac{2}{3}J_1^2(k) - 2J_0(k)J_2(k) - \tfrac{1}{3}J_2^2(k)\,. \tag{45d}$$

Finally,

$$(g_3,g_2) = -\frac{2i}{\pi^2}\int_{-1}^{1}(1-x^2)\,dx\int_{x}^{1}e^{-i\sigma(x-\eta)}\frac{g_1(\eta)\,d\eta}{(1-\eta^2)^{\frac{1}{2}}}\int_{-1}^{1}\frac{e^{ik\xi}\,d\xi}{(1-\xi^2)^{\frac{1}{2}}(\xi-x)}$$

$$= \sum_{m=1}^{\infty}\sum_{n=1}^{\infty}\frac{1}{m}[\,N_m(\sigma) - (i)^m N_0(\sigma)]\,J_n(k)\{(-1)^n[\,J_{m-n-2}(\sigma) + 2J_{m-n}(\sigma)$$

$$+ J_{m-n+2}(\sigma)] - [\,J_{m+n-2}(\sigma) + 2J_{m+n}(\sigma) + J_{m+n+2}(\sigma)]\}\,, \tag{45e}$$

where

$$N_n(\sigma) = -[1-\Theta(\sigma)]\,J_n'(\sigma) - i\Theta(\sigma)\,J_n(\sigma) \qquad (n=0,1,2,\ldots\ldots). \tag{45f}$$

The above result can be shown by using the series expansion of g_2 and g_3 as

$$g_2(x) = 2i\sin\theta \sum_{n=1}^{\infty} (-i)^n J_n(k)\sin n\theta \qquad (x=\cos\theta)\ ,$$

$$g_3(x) = 2ie^{-i\sigma\cos\theta}\sin\theta \sum_{n=1}^{\infty} (i)^n[\,N_n(\sigma) - (i)^n N_0(\sigma)]\,\frac{\sin n\theta}{n} \qquad (x=\cos\theta)\ .$$

This completes the determination of a_3, hence also the orthogonalization.

It is now evident that v can be expressed as

$$v(x) = \sum_{n=1}^{3} B_n f_n(x) + v_\perp(x)\ , \qquad (-1\leqslant x\leqslant 1)\ , \tag{46}$$

where B_n's are complex coefficients and $v_\perp$ is any function belonging to the orthogonal complement of the subspace spanned by $\{f_1, f_2, f_3\}$, that is, $<f_n, v_\perp> = 0$ for $n=1,2,3$. For convenience of the subsequent computations, we introduce the real parameters ξ_n's by

$$<v, f_n> = B_n <f_n, f_n> \equiv \xi_{2n-1} + i\xi_{2n} \qquad (n=1,2)\ , \tag{47a}$$

$$<v, f_3> = B_3 <f_3, f_3> \equiv \xi_5 + i\xi_6 - (C_1 + iC_2)\ , \tag{47b}$$

where $C_1 + iC_2$ is given by (42c). From (43), (46) and (47), we have

$$<v, g_2> = a_1^* <v, f_1> + <v, f_2> = a_1^*(\xi_1 + i\xi_2) + (\xi_3 + i\xi_4)\ , \tag{48a}$$

$$<v, g_3> = a_2^*(\xi_1 + i\xi_2) + a_3^*(\xi_3 + i\xi_4) + (\xi_5 + i\xi_6) - (C_1 + iC_2)\ . \tag{48b}$$

Substitution of (47), (48) in (33)$'$ and (34) yields

$$C_P = \sigma\{A_2(\sigma)(\xi_1^2+\xi_2^2) + A_3(\xi_2\xi_3 - \xi_1\xi_4) + A_4(\xi_1\xi_3 + \xi_2\xi_4) + (\xi_2\xi_5 - \xi_1\xi_6)$$
$$+ 2\epsilon\,[\sum_{j=1}^{4} P_j\xi_j - J_0(k)\xi_5 + J_1(k)\xi_6]\}\ , \tag{49}$$

$$C_E = B(\sigma)(\xi_1^2+\xi_2^2) + 2\epsilon\,(Q_1\xi_1 + Q_2\xi_2 + Q_3\xi_3) - 4\epsilon^2 W^2\ , \tag{50}$$

where

$$a_2 = A_1(\sigma) + iA_2(\sigma)\ , \qquad a_3 = A_3(\sigma,k) + iA_4(\sigma,k)\ , \tag{51a}$$

$$P_1 = -A_1J_0(k) - A_2J_1(k) + \frac{2}{3k}J_1(k)\ , \quad P_2 = A_1J_1(k) - A_2J_0(k) - J_2(k)/3k\ , \tag{51b}$$

$$P_3 = -A_3 J_0(k) - A_4 J_1(k) + 1/k \ , \qquad P_4 = A_3 J_1(k) - A_4 J_0(k) \ , \tag{51c}$$

$$Q_1 = (1 - 2\mathfrak{F}) W_1 + 2\mathfrak{G} W_2 + \tfrac{2}{3}(\tfrac{\sigma}{k} - 1) J_1 \ , \tag{51d}$$

$$Q_2 = 2\mathfrak{G} W_1 - (1 - 2\mathfrak{F}) W_2 - \tfrac{1}{3}(\tfrac{\sigma}{k} - 1) J_2 \ , \quad Q_3 = (\sigma/k) - 1 \ , \tag{51e}$$

The other coefficients appeared here have been given in (35), (45).

Equations (49), (50) show that C_P depends on only six real parameters $\{\xi_1, \xi_2, \ldots \xi_6\}$, and C_E depends on only three parameters $\{\xi_1, \xi_2, \xi_3\}$, while both C_P and C_E, hence also C_T, are independent of $v_\perp(x)$. (Note that the orthogonal complement of the subspace spanned by $\{f_1, f_2, f_3\}$ is infinite dimensional.) Thus, it is clear that the optimization problem posed earlier now reduces to one defined on a finite-dimensional vector space.

Before we proceed with our discussion from this approach, further simplification of the expressions for C_P and C_E can be gained if we first eliminate the terms linear in ξ_1 and ξ_2 in (49), (50) and then reduce the number of quadratic terms in (49) by the following transformation

$$\zeta_1 + i\zeta_2 = \frac{1}{A}(A_4 - iA_3)[\xi_1 + i\xi_2 + \frac{\epsilon}{B}(Q_1 + iQ_2)] \ , \quad A = (A_3^2 + A_4^2)^{\frac{1}{2}} \ , \tag{52a}$$

$$\zeta_3 + i\zeta_4 = (\xi_3 + i\xi_4) + (A_3 + iA_4)(\zeta_5 + i\zeta_6)/A^2 \ , \tag{52b}$$

$$\zeta_5 + i\zeta_6 = (\xi_5 + i\xi_6) + (C_5 + iC_6) \ , \tag{52c}$$

where

$$C_5 = 2\epsilon\,(P_2 - A_2 Q_2/B) \ , \qquad C_6 = -2\epsilon\,(P_1 - A_2 Q_1/B) \ . \tag{52d}$$

Then (49) and (50) reduce to

$$C_P/\sigma = A_2(\zeta_1^2 + \zeta_2^2) + A(\zeta_1\zeta_3 + \zeta_2\zeta_4) + \epsilon\,(A_5\zeta_3 + A_6\zeta_4) + \epsilon\,A_0 \tag{53}$$

$$C_E = B(\zeta_1^2 + \zeta_2^2) + 2\epsilon\,Q_3\zeta_3 - \epsilon\,Q_0 \ , \tag{54}$$

where

$$A_5 = 2P_3 - (A_3 Q_2 + A_4 Q_1)/B \ , \qquad A_6 = 2P_4 - (A_4 Q_2 - A_3 Q_1)/B \ , \tag{55a}$$

$$\begin{aligned} A_0 = & -2[J_0 + (P_3 A_3 + P_4 A_4)/A^2]\,\zeta_5 + 2[J_1 + (P_3 A_4 - P_4 A_3)/A^2]\,\zeta_6 \\ & + 4\epsilon\,[J_0 P_2 + J_1 P_1 - A_2(J_0 Q_2 + J_1 Q_1)/B] - \epsilon\,A_2(Q_1^2 + Q_2^2)/B^2 \ , \end{aligned} \tag{55b}$$

$$Q_0 = 2Q_3(A_3\zeta_5 - A_4\zeta_6)/A^2 + 4\epsilon\,W^2 + \epsilon\,(Q_1^2 + Q_2^2)/B \ . \tag{55c}$$

Thus in the above reduced form, C_P depends quadratically on $\{\zeta_1, \zeta_2, \zeta_3, \zeta_4\}$, C_E

depends quadratically on $\{\zeta_1, \zeta_2\}$ but is independent of ζ_4, while both C_P and C_E depend linearly on $\{\zeta_5, \zeta_6\}$. When the primary wave is absent (i.e. $\epsilon = 0$), equations (53) and (54), or equivalently (49) and (50), reduce to the case of a flexible plate in uniform flow treated earlier by Wu (1971b, see his equations (79) and (80), which involve also six independent parameters). The present result of C_P and C_E is also very similar to that for a flat plate oscillating in waves discussed by Wu (1972, see his equations (50), (51) for the four independent parameters proper to that problem). Like those simpler cases investigated previously, we note that in the three-dimensional Euclidean space $(\zeta_1, \zeta_2, \zeta_3)$ (i.e. with $\zeta_4, \zeta_5, \zeta_6$ held fixed), the C_E= const. surfaces are paraboloids of revolution with its generating axis lying along the ζ_3-axis, while C_P= const. surfaces are oblique hyperboloids, whose cross-sections with ζ_3=const. planes, if real, are circles.

The optimization problem posed earlier can now be reformulated as follows: Let R_6 denote the six-dimensional Euclidean space of ordered six-tuples $\vec{\zeta} \equiv (\zeta_1, \zeta_2, \zeta_3, \zeta_4, \zeta_5, \zeta_6)$ of real numbers; and let Ω be a subset of R_6 defined by

$$\Omega \equiv \{ \vec{\zeta} \in R_6 : \quad C_T(\vec{\zeta}) = C_P(\vec{\zeta}) - C_E(\vec{\zeta}) = C_{T,o} > 0 \} . \tag{56}$$

The optimization problem is to find a vector $\vec{\zeta}^{\,o} \in \Omega$ such that $C_E(\vec{\zeta}^{\,o}) \leqslant C_E(\vec{\zeta})$ for all $\vec{\zeta} \in \Omega$.

From the known geometric properties of constant C_P and C_E surfaces, and hence also of $C_T = C_{T,o} > 0$ surface, it follows that Ω is an unbounded set in R_6. Consequently, it is possible that the optimization problem may not have a solution. It suffices to demonstrate two such cases. As the first, consider a sequence of points $\{\vec{\zeta}^{\,k}\}$ in the set $S_1 \subset \Omega$ defined by

$$S_1 = \{ \vec{\zeta} \in \Omega : \quad \sigma A \zeta_1 + \epsilon\,(\sigma A_5 - 2Q_3) = 0 \} , \tag{57}$$

such that $Q_3 \zeta_3^k \to -\infty$ as $k \to \infty$. It is readily shown that in the set S_1, C_T depends on $(\zeta_2, \zeta_4, \zeta_5, \zeta_6)$ while C_E depends on $(\zeta_2, \zeta_3, \zeta_5, \zeta_6)$; S_1 is therefore nonempty. But, since ζ_j^k's are all constant for j=2, 4, 5, 6 and for all k, we immediately see from (54) that the sequence of values $C_E(\vec{\zeta}^{\,k}) \to -\infty$, as $k \to \infty$, implying the nonexistence of an optimal solution.

As the second example, consider another sequence of points $\{\vec{\zeta}^{\,\ell}\}$ in the set $S_2 \subset \Omega$ defined by

$$S_2 = \{ \vec{\zeta} \in \Omega : \quad \sigma A_o(\zeta_5, \zeta_6) + Q_o(\zeta_5, \zeta_6) = 0 \} , \tag{58}$$

such that $Q_o(\zeta_5^\ell, \zeta_6^\ell) \to \infty$ as $\ell \to \infty$. It is also easily seen that in the set S_2, $C_T = C_T(\zeta_1, \zeta_2, \zeta_3, \zeta_4)$, $C_E = C_E(\zeta_1, \zeta_2, \zeta_3, \zeta_5)$ and consequently $C_E(\vec{\zeta}^{\,\ell}) \to -\infty$ while C_T remains unchanged as $\ell \to \infty$, implying again the nonexistence of an optimal solution.

To ensure the existence of an optimal solution that is physically meaningful, we shall minimize C_E over a subset $\tilde{\Omega}$ of Ω which is closed and bounded, i.e.

$$\tilde{\Omega} \equiv \{ \vec{\zeta} \in \tilde{R}_6 \ : \quad C_T(\vec{\zeta}) = C_P(\vec{\zeta}) - C_E(\vec{\zeta}) = C_{T,o} > 0 \} , \tag{59a}$$

where $\tilde{R}_6$ denote a bounded subset of R_6 such that

$$\sum_{n=1}^{6} \zeta_n^2 \leqslant M < \infty . \tag{59b}$$

The new optimization problem is to find a vector $\vec{\zeta}^o \in \tilde{\Omega}$ such that $C_E(\vec{\zeta}^o) \leqslant C_E(\vec{\zeta})$ for all $\vec{\zeta} \in \tilde{\Omega}$. Evidently, this optimization problem has a solution since C_E is continuous on the closed bounded set $\tilde{\Omega}$.

In what follows, we shall consider the particular case where ζ_4, ζ_5 and ζ_6 are treated as free parameters so that the optimization problem reduces to a three-dimensional one. Moreover, the constant M in (59b) is adjusted so that the optimum solution can be determined from the points in $\tilde{\Omega}$ at which (grad C_P) is proportional to (grad C_E). Thus, we set

$$\partial\, (C_P - \lambda' \sigma C_E) / \partial \zeta_j = 0 \ , \qquad j=1,2,3, \tag{60}$$

where λ' is a Lagrange multiplier, giving

$$\zeta_1 = \lambda A \zeta_3 \ , \tag{61a}$$

$$\zeta_2 = \lambda A \zeta_4 \ , \tag{61b}$$

$$\zeta_1 = (\epsilon / AB)(2A_2Q_3 - A_5B) + (\epsilon\, Q_3/AB)\, \lambda^{-1} \ , \tag{61c}$$

where λ is related to λ' by $\lambda^{-1} = 2(B\lambda' - A_2)$. From the three equations (61a-c) we can determine the variables $(\zeta_1, \zeta_2, \zeta_3)$, which are subject to variation, in terms of $(\zeta_4, \zeta_5, \zeta_6, \lambda)$. Finally, the Lagrange multiplier λ can be determined in terms of ζ_4, ζ_5, ζ_6, $C_{T,o}$ and ϵ by invoking condition (39). This line of approach indicates that the extremal solution will involve $(\zeta_4, \zeta_5, \zeta_6, C_{T,o}, \epsilon)$ as free parameters. It is more desirable, however, to adopt

$$\zeta_o = (\zeta_3^2 + \zeta_4^2)^{\frac{1}{2}} \tag{62}$$

rather than ζ_4 as a free parameter since this replacement will facilitate computation as well as comparison with the earlier results for the uniform stream case (Wu, 1971b) and those for the rigid plate in waves (Wu, 1972). Thus, we first eliminate ζ_1, ζ_2 in (53), (54), and (61a-c), next we apply condition (39), giving

$$A^2 [\, T_2 \lambda^2 + \sigma \lambda] + \bar{\epsilon} [\, (\sigma A_5 - 2Q_3) z_3 + \sigma A_6 z_4] = \bar{C}_{T,o} - \bar{\epsilon}\, \Sigma_o \ , \tag{63}$$

$$z_3 = (\bar{\epsilon}/A^2B)(2A_2Q_3 - A_5B)\frac{1}{\lambda} + (\bar{\epsilon}Q_3/A^2B)\frac{1}{\lambda^2} \quad , \tag{64}$$

$$z_4 = \pm(1 - z_3^2)^{\frac{1}{2}} \quad , \tag{65}$$

where

$$T_2 = \sigma A_2 - B \quad , \qquad z_j = \zeta_j/\zeta_0 \quad (j = 1, 2, \ldots 6) \quad , \tag{66a}$$

$$\bar{C}_{T,o} = C_{T,o}/\zeta_0^2 \quad , \quad \bar{\epsilon} = \epsilon/\zeta_0 \quad , \quad \Sigma_0 = (\sigma A_0 + Q_0)/\zeta_0 \quad . \tag{66b}$$

Equation (65) follows from the definition of z_3, z_4 and ζ_0 as given by (62) and (66a), there being two branches of z_4 for given z_3, with $|z_3| \leqslant 1$.

The three equations (63)-(65) involve three unknowns, z_3, z_4, λ, and three parameters, namely $\bar{C}_{T,o}$ — the "proportional loading factor", $\bar{\epsilon}$ — the "proportional wave factor", and Σ_0 — the "complementary mode factor" which includes the contribution from the mode $\zeta_5 + i\zeta_6$ and that from the waves.

As for the actual calculation of the Lagrange multiplier λ, we note that if equations (64), (65) are substituted for z_3, z_4 in equation (63), then, upon perfect squaring, there results an eighth degree algebraic equation for λ, of which the real solutions (appearing always in even numbers) are of interest. This equation seems to be too difficult for analytical solutions; resort was then made to numerical methods. The method which proved to be successful is as follows. Since the physically meaningful solutions also require z_3, z_4 to be both real, equation (65) suggests that z_3 can be used effectively for parametric computation of λ, with the obvious advantage of having a bounded range $-1 \leqslant z_3 \leqslant 1$. In this parametric form, both equations (63) and (64) are quadratic equations in λ, each giving two solutions of λ in closed form, from which the real solutions of λ satisfying both equations were determined by Newton's method, using z_3 as a parameter. The number of solutions depend on the values of σ, $\bar{C}_{T,o}$, $\bar{\epsilon}$ and Σ_0 ; on occasions as many as eight real solutions were obtained, and there are cases in which two real solutions are very close to each other. In all the cases tried the two real solutions providing the highest and lowest efficiencies were taken as the desired optimal solutions.

The numerical results of η_{max}, as shown in Figs 2 - 5 for a few representative cases, exhibit the following salient features of the optimum solution. For $\bar{C}_{T,o}$ and $\bar{\epsilon}$ both as small as 10^{-3} and with $z_5 = z_6 = 0$, the maximum efficiency is already $\eta_{max} = 1.0$ for the reduced frequency $\sigma > 10^{-2}$. The corresponding results of the maximum efficiency for a rigid hydrofoil in a uniform stream ($\bar{\epsilon} = 0$) and in regular waves ($\bar{\epsilon} > 0$) are reproduced (see Wu 1971b, 1972) in Fig. 3 over a similar range of $\bar{C}_{T,o}$, although the $\bar{C}_{T,o}$ and $\bar{\epsilon}$ in those two cases are defined with reference to the heaving amplitude at mid-chord, whereas the definition of $\bar{C}_{T,o}$ and $\bar{\epsilon}$ in the present case (by referring to ζ_0, see (66), whose physical significance is not quite so simple) is slightly different. With this qualification, a comparison between Fig. 2 and 3 shows that η_{max} is further improved by the flexible over the rigid foil, in the frequency range of interest. When $\bar{C}_{T,o}$ is kept at 10^{-3} and $\bar{\epsilon}$ alone is increased (by having a stronger wave) to 10^{-2}, η_{max} becomes greater than 1, corresponding to the operation in which energy is extracted from the surrounding waves, but a power (somewhat smaller than before) is still required for maintaining the hydrofoil motion. When $\bar{\epsilon}$ is as large as 0.1, we see that η_{max} becomes negative, indicating that both energy and power are supplied by the exterior wave field. At this high level of $\bar{\epsilon}$, η_{max} becomes more negative as $\bar{C}_{T,o}$ is increased to 10^{-2}. This trend implies that more energy and power can be extracted from stronger waves at higher loadings. This is a quite remarkable feature since this trend is reversed from that at smaller $\bar{\epsilon}$ (or weaker waves, see the curves with $\bar{\epsilon} = 10^{-2}$).

To summarize the case of $z_5 = z_6 = 0$, we note that the overall features of η_{max} of the flexible and rigid hydrofoils are very similar, the difference, after the two definitions of $\bar{C}_{T,o}$ and $\bar{\epsilon}$ are properly reconciled, being rather small. Since the salient features of the optimum motion, including the variation of the leading-edge suction C_S, feathering of the hydrofoil to the trajectory, etc., have been thoroughly explored for the rigid plate case (see Wu, 1972), these features will not be further pursued here for the flexible plate. However, it must be stressed that the additional degrees of freedom provided by z_5 and z_6 for flexible plates can alter the η_{max} of a flexible plate. As indicated by Figs. 4 and 5 for two typical examples, η_{max} for the basic case of $z_5 = z_6 = 0$ can be increased by suitable choice of z_5 and z_6 (within the set $\tilde{R}_6$ of (59)). The trend of the influence by z_5 and z_6 on the value of η_{max} can be seen clearly through the linear dependence of Σ_o on z_5 and z_6 (see (63) - (66)). Obviously, η_{max} will remain unchanged when the set of parameters ($\bar{C}_{T,o}$, $\bar{\epsilon}$, z_5, z_6) is replaced by ($\bar{C}'_{T,o}$, $\bar{\epsilon}$, 0, 0), where

$$\bar{C}'_{T,o} = \bar{C}_{T,o} - \bar{\epsilon}\,\Sigma_o(z_5, z_6) \; . \tag{67}$$

This explains the opposite trend of η_{max} from its basic value at $z_5 = z_6 = 0$ when a set of nonvanishing z_5 and z_6 is reversed in sign. By making use of this property, or equivalently the simple formula (67), the utility and interpretation of Fig. 2 is thereby greatly extended.

Acknowledgments

This work was partially sponsored by the National Science Foundation, under Grant GK 10216, and by the Office of Naval Research, under Contract N00014-67-A0094-0012.

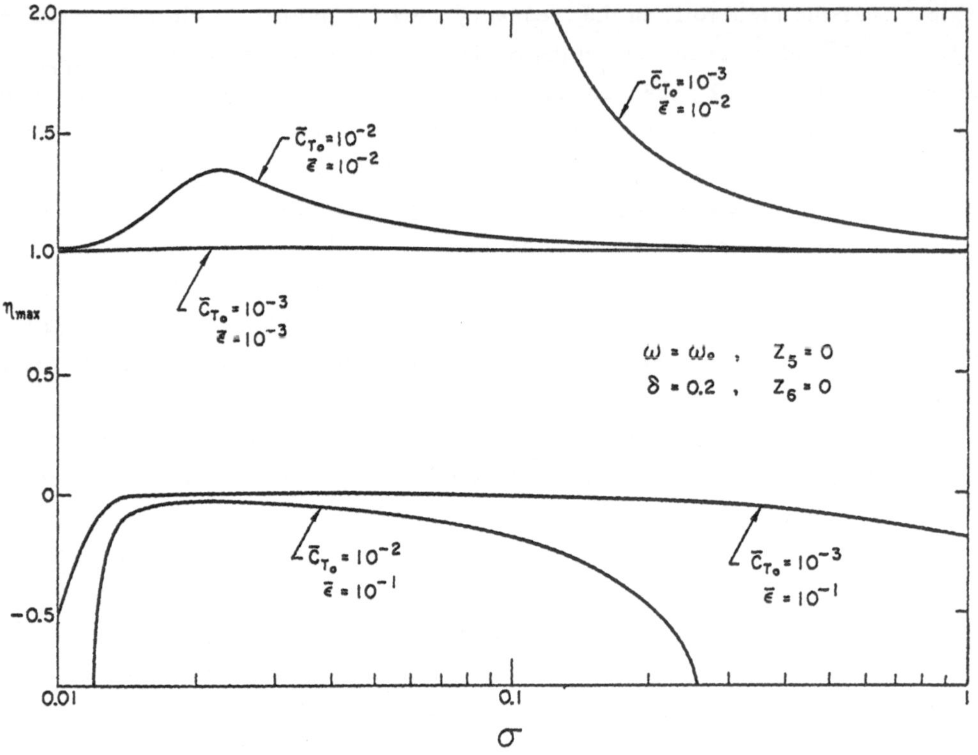

2.0
1.5
1.0
0.5
0
-0.5
η_{max}
0.01
0.1
1
σ
$\bar{C}_{T_0} = 10^{-3}$
$\bar{\epsilon} = 10^{-2}$
$\bar{C}_{T_0} = 10^{-2}$
$\bar{\epsilon} = 10^{-2}$
$\bar{C}_{T_0} = 10^{-3}$
$\bar{\epsilon} = 10^{-3}$
$\omega = \omega_0$, $Z_5 = 0$
$\delta = 0.2$, $Z_6 = 0$
$\bar{C}_{T_0} = 10^{-2}$
$\bar{\epsilon} = 10^{-1}$
$\bar{C}_{T_0} = 10^{-3}$
$\bar{\epsilon} = 10^{-1}$

FIGURE 2

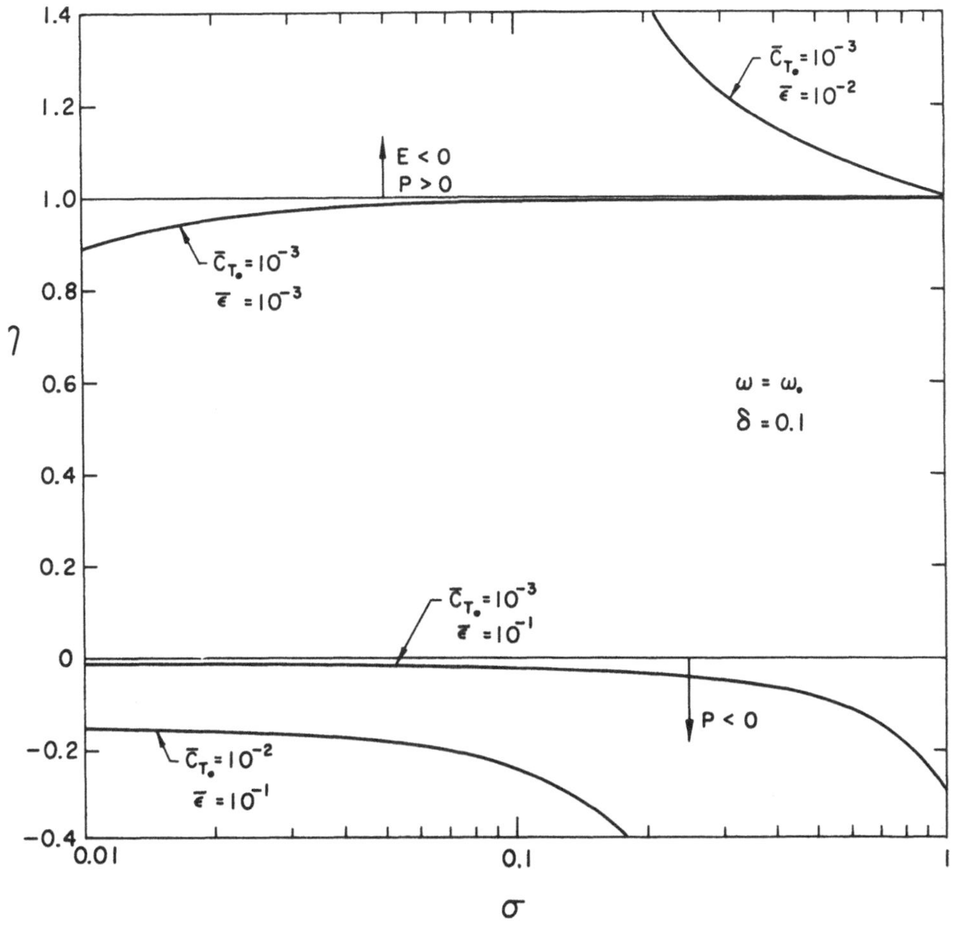

1.4
1.2
1.0
0.8
0.6
0.4
0.2
0
−0.2
−0.4
0.01
0.1
1
σ
E < 0
P > 0
P < 0
ω = ω₀
δ = 0.1
$\bar{C}_{T_0}=10^{-3}$
$\bar{\epsilon}=10^{-2}$
$\bar{C}_{T_0}=10^{-3}$
$\bar{\epsilon}=10^{-3}$
$\bar{C}_{T_0}=10^{-3}$
$\bar{\epsilon}=10^{-1}$
$\bar{C}_{T_0}=10^{-2}$
$\bar{\epsilon}=10^{-1}$

FIGURE 3

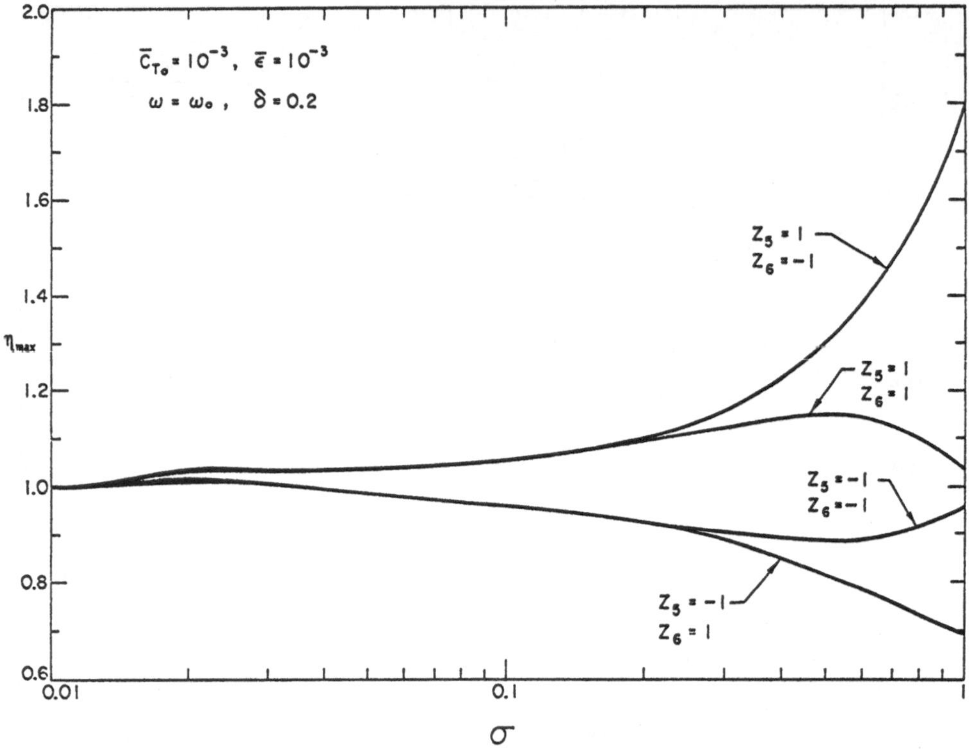

$\bar{C}_{T_0} = 10^{-3}$, $\bar{\epsilon} = 10^{-3}$
$\omega = \omega_0$, $\delta = 0.2$
$Z_5 = 1$ $Z_6 = -1$
$Z_5 = 1$ $Z_6 = 1$
$Z_5 = -1$ $Z_6 = -1$
$Z_5 = -1$ $Z_6 = 1$
η_{max}
2.0
1.8
1.6
1.4
1.2
1.0
0.8
0.6
0.01
0.1
1
σ

FIGURE 4

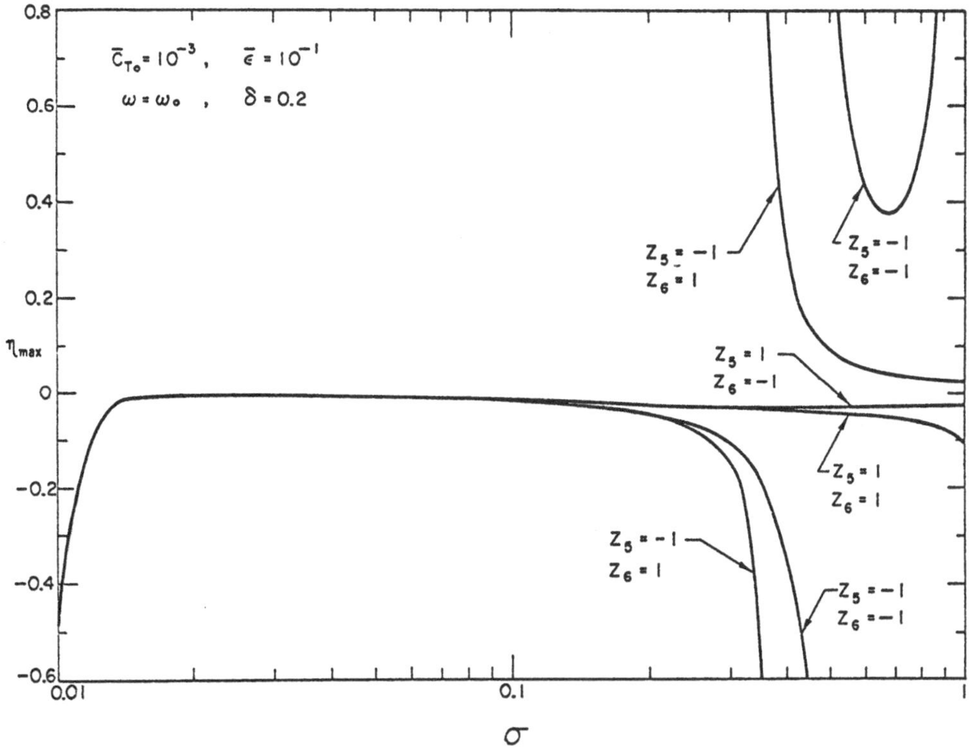
$\bar{C}_{T_0} = 10^{-3}$, $\bar{\epsilon} = 10^{-1}$
$\omega = \omega_0$, $\delta = 0.2$
η_{max}
σ
0.8
0.6
0.4
0.2
0
-0.2
-0.4
-0.6
0.01
0.1
1
$Z_5 = -1$ $Z_6 = 1$
$Z_5 = -1$ $Z_6 = -1$
$Z_5 = 1$ $Z_6 = -1$
$Z_5 = 1$ $Z_6 = 1$
$Z_5 = -1$ $Z_6 = 1$
$Z_5 = -1$ $Z_6 = -1$

FIGURE 5

References

Garrick, I. E. 1936 Propulsion of a flapping and oscillating airfoil. NACA TR 567.

Garrick, I. E. 1957 Nonsteady Wing Characteristics. Sect. F., High Speed Aerodynamics and Jet Propulsion, Vol. 7 (ed. A. F. Donovan and H. R. Lawrence), Princeton University Press, Princeton, New Jersey.

Küssner, H. G. 1935 Augenblicklicher Entwicklungsstand der Frage des Flügelflatterns. Luftfahrtforschung, Vol. 6, 193-209.

Küssner, H. G. 1940 Das Zweidimensionale Problem der beliebig bewegten Tragflache unter Berücksichtigung von partialbewegungen der Flussigkeit. Luftfahrtforschung, Vol. 17, 355-361.

Lighthill, M. J. 1960 Note on the swimming of slender fish. J. Fluid Mech. Vol. 9, 305-317.

Lighthill, M. J. 1969 Hydromechanics of aquatic animal propulsion. Annu. Rev. Fluid Mech. Vol. 1, 413-445.

Lighthill, M. J. 1970 Aquatic animal propulsion of high hydromechanical efficiency. J. Fluid Mech. Vol. 44, 265-301.

Luke, Y. L. and Dengler, M. A. 1951 Tables of the Theodorsen circulation function for generalized motion. J. Aero. Sci. Vol. 18, 478-483.

Osborne, M. F. M. 1960 The hydrodynamical performance of migratory salmon. J. Exp. Biol. Vol. 38, 365-390.

Sears, W. R. 1940 Some aspects of nonstationary airfoil theory and its practical application. J. Aero. Sci. Vol. 8, 104-108.

von Kármán, Th. and Sears, W. R. 1938 Airfoil theory for non-uniform motion. J. Aero. Sci. Vol. 5, 379-390.

Weinblum, G. P. 1954 Approximate theory of heaving and pitching of hydrofoils in regular shallow waves. DTMB Report C-479.

Wu, T. Y. 1954 A theory for hydrofoils of finite span. J. Math. Phys. Vol. 33, 207-248.

Wu, T. Y. 1961 Swimming of a waving plate. J. Fluid Mech. Vol. 10, 321-344.

Wu, T. Y. 1971a Hydromechanics of swimming propulsion. Part 1. Swimming of a two-dimensional flexible plate at variable forward speeds in an inviscid fluid. J. Fluid Mech. Vol. 46, 337-355.

Wu, T. Y. 1971b Hydromechanics of swimming propulsion. Part 2. Some optimum shape problems. J. Fluid Mech. Vol. 46, 521-544.

Wu, T. Y. 1972 Extraction of flow energy by a wing oscillating in waves. J. Ship Res. Vol. 14, No. 1, 66-78.

STABILITY THEORY FOR GENERAL DYNAMICAL SYSTEMS AND SOME APPLICATIONS

E. F. Infante

Center for Dynamical Systems
Division of Applied Mathematics
Brown University
Providence, Rhode Island

In this lecture, it will be attempted to describe some recent results in the stability of general dynamical systems from the viewpoint of Liapunov theory, and to illustrate and motivate them through examples.

The subject of stability theory is an extremely broad one. During the last ten years a large number of developments have taken place in this area; it could be said that, beside Liapunov theory and the classical theory for finite dimensional systems, the new branch of the functional analytic approach has sprung during this period.

This lecture is purposefully limited to a description of the Liapunov approach, and a rather limited one at that. To attempt more would surely lead to failure; furthermore, the author has worked much more extensively in this area than in the other ones. Fortunately, some recent expositions of a very readable nature are available [4, 13] to those interested in the functional analytic approach; furthermore, at the foundations of the functional analytic approach is the concept of a dynamical system, the central mathematical concept of this lecture.

Before embarking on the subject, the author feels obligated to apologize to the numerous workers in this area which he will neither reference nor acknowledge contributions from. This, it is hoped, is permissible, given the tutorial nature of the lecture. Interested readers will find the appropriate references in the few papers quoted.

The development of stability theory, from the time of Poincare and Liapunov, was considered a branch of ordinary differential equations and of mechanics; interest in the stability theory of partial differential equations and functional

---

This research was supported by the Office of Naval Research, NONR N0014-67-A-0191-0009

differential equations is rather more recent, especially in the engineering literature. The fundamental idea behind the concept of a dynamical system is to try to generalize to broader classes of evolutions the results known for ordinary differential equations.

1. Dynamical Systems and Some Examples

Let R^n denote an n-dimensional vector space with norm $|\cdot|$, R^+ denote the interval $[0,\infty)$ and $\mathcal{B}$ a Banach space with $\|\varphi\|$ the norm of an element in $\mathcal{B}$. Then [9, 11]

Definition 1.1. A dynamical system in a Banach space $\mathcal{B}$ is a function $u: R^+ \times \mathcal{B} \to \mathcal{B}$ such that

(i) u is continuous

(ii) $u(0,\varphi) = \varphi$

(iii) $u(t+\tau,\varphi) = u(t,u(\tau,\varphi))$ for every $t,\tau \geq 0$, every φ in $\mathcal{B}$.

Hence, a dynamical system has some continuity properties, the second condition states that at $t = 0$ the dynamical system is the identity map and, finally, that it has the semigroup property. It will be noted that the definition implies that a dynamical system is autonomous and that the map of the dynamical system is defined only forward in time. Except for this last restriction, it represents, at a slightly more abstract level, precisely the properties associated with the solutions of ordinary differential equations of autonomous type. Associated with the dynamical system we have

Definition 1.2. The positive orbit $O^+(\varphi)$ through φ is the set of elements in $\mathcal{B}$ defined by $O^+(\varphi) = \bigcup_{t\geq 0} u(t,\varphi)$.

Let us give some examples of dynamical systems.

Example 1. Ordinary differential equations. Consider the equation

$$\dot{x} = f(x), \tag{1.1}$$

where $f: R^n \to R^n$, is continuous for every ξ in R^n, and the solution $u(t,\xi)$, $u(0,\xi) = \xi$ exists for all $t \geq 0$, is unique and depends continuously upon t,ξ. Uniqueness of the solution implies $u(t+\tau,\xi) = u(t,u(\tau,\xi))$ for all t,τ and therefore for all $t,\tau \geq 0$. Clearly u is a dynamical system on R^n.

<u>Example 2.</u> <u>Functional differential equations of the retarded type.</u> Let $C = C([-r,0],R^n)$, $r \geq 0$, be the space of continuous functions from $[-r,0]$ to R^n with the uniform convergence topology. For any continuous function x defined on $[-r,s)$, $s > 0$ and any $0 \leq t < s$ let x_t be the function in C defined by $x_t(\theta) = x(t+\theta)$, $-r \leq \theta \leq 0$. Then the functional differential equation

$$\dot{x}(t) = f(x_t), \tag{1.2}$$

with $f: C \to R^n$ continuous and locally lipschitz will have a solution $x = x(\varphi)(t)$ defined and continuous on $[-r,s)$, $s > 0$ and $x_0 = \varphi$, the initial value for every φ in C. With $u(t,\varphi) = x_t(\varphi)$, since local existence, uniqueness and continuous dependence is easily proved [12] then u is a dynamical system on C.

It should be noted that this functional differential equation (simplest example $\dot{x}(t) = \alpha x(t-1)$) only defines solutions forward in time, hence the dynamical system definition.

<u>Example 3.</u> <u>Functional differential equations of the neutral type.</u> Let $C = C([-r,0),R^n)$ with the same norm as above and, with the same notation consider the equation

$$\frac{d}{dt}(Dx_t) = f(x_t) \tag{1.3}$$

where D is a difference operator defined by

$$D\varphi = \varphi(0) - \sum_{k=1}^{N} A_k\varphi(-\tau_k) \tag{1.4}$$

where A_k are $n \times n$ constant matrices and $0 < \tau_k \leq r$ with τ_j/τ_n rational. This is a special case of a functional differential equation of the neutral type (for example, $\dot{x}(t) + d\dot{x}(t-1) + bx(t) + cx(t-1) = 0$ is such an equation). If f is continuous and locally lipschitzian in C then it is possible to show that with any initial value φ in C the solution $u(t,\varphi) = x(\varphi)$ exists, is continuous in t and φ and is unique [10]. If solutions exist for all $t \geq 0$, then $u(t,\varphi)$ defines a dynamical system on C. Note again that the solution can be defined only forward in time.

Example 4. Parabolic partial differential equations. In this case, consider the heat equation, with boundary and initial conditions

$$\begin{aligned} &u_t = u_{xx}, \quad 0 \leq x \leq \pi, \quad t > 0 \\ &u(0,t) = u(\pi,t) = 0, \quad t \geq 0 \\ &u(x,0) = \phi(x) \end{aligned} \tag{1.5}$$

Consider the space X of functions $\phi\colon [0,\pi] \to R$ continuously differentiable on $[0,\pi]$ with $\phi(0) = \phi(\pi) = 0$ and with norm $\|\phi\|_1 = \sup\{|\phi'(x)|: 0 \leq x \leq \pi\}$. Then it is well-known that $u(t,\phi) = u(t,x;\phi(x))$ exists for all $t \geq 0$ is unique and depends continuously on t,ϕ, in the norm of X. Hence, we have a dynamical system in the Banach space X. Note that, once again, the solutions are not defined backward in time - as is well-known the backward solution will not be in X.

Example 5. Consider the equation

$$\begin{aligned} &v_{tt} = v_{xx} + f(v,v_t,v_x) \quad 0 \leq x \leq 1, \ t \geq 0 \\ &v(0,x) = \varphi(x), v_t(0,x) = \psi(x) \\ &v(t,0) = 0, \quad v(t,1) = 0, \quad t \geq 0 \end{aligned} \tag{1.6}$$

where f is analytic in its variables in the whole space. Let W_2^k the space of functions with all generalized derivatives of order less than an equal to k square integrable in $[0,1]$ with norm $\|\varphi\|^2_{W_2^k} = \int_0^1 [\varphi^2 + (\varphi^1)^2 + \cdots + (\varphi^{(k)})^2]dx$, where $\varphi^{(j)}$ is the jth generalized derivative of φ. Then [19], it is known that (1.6) has a unique generalized solution $v(t,x,\varphi,\psi)$ on $-\eta \leq t \leq \eta$ for every φ in W_2^k and any ψ in W_2^{k-1} and that the pair $[v(t,x; \varphi,\psi), v_t(t,x; \varphi,\psi)]$ belongs to $W_2^k \times W_2^{k-1}$ and is continuous in t,φ,ψ. Hence, if it is assumed that such a solution exists for all $t \geq 0$, then $u(t,\Phi) = [(v(t,x; \varphi,\psi), v_t(t,x; \varphi,\psi)]$, is a dynamical system on $W_2^k \times W_2^{k-1}$ for any $k \geq 1$.

The purpose of these examples has been to illustrate the generality of the concept of dynamical systems. We shall return to some specific applications of a physical nature later.

2. Some Stability Theorems

Let us now state, for our general dynamical system, the fundamental theorems which we wish to exploit for the determination of stability results. For this purpose, let

Definition 2.1. Let a dynamical system $u(t,\varphi)$ be defined in the Banach space $\mathcal{B}$. If $u(t,\psi) = \psi$ for all $t \geq 0$, then ψ is an equilibrium solution of the dynamical system.

Definition 2.2. The equilibrium solution $\varphi = 0$ of $u(t,\varphi)$ is stable, if for every $\epsilon > 0$ there exists a $\delta(\varepsilon)$ such that $\|\psi\| \leq \delta$ implies $\|u(t,\psi)\| \leq \epsilon$ for all $t \geq 0$. The equilibrium $\varphi = 0$ is asymptotically stable if it is stable and there exists a γ such that $\|\psi\| \leq \gamma$ implies $\lim_{t \to \infty} u(t,\psi) \to 0$ (in the norm in $\mathcal{B}$).

Definition 2.3. A set M in $\mathcal{B}$ is a positively invariant set of the dynamical system u if for each Φ in M, $O^+(\Phi) \subset M$. It is invariant if for each Φ in M

there exists a function $U(s,\Phi)$, $U(0,\Phi) = \Phi$ defined and in M for $-\infty < s < \infty$ and such that $u(t,U(s,\Phi)) = U(t+s,\Phi)$ for all $t \geq 0$.

Definitions 2.1 and 2.2 are the natural generalization of the familiar ones. The first part of Definition 2.3 is well-known; the second part of the definition simply uses the device of extending the dynamical system backward, if possible, since the dynamical system is not defined backward. Note that the function U must exist only for those Φ in M.

Let us now define, in the manner of [9, 11]

Definition 2.4. If u is a dynamical system on $\mathscr{B}$ and V is a continuous scalar function on $\mathscr{B}$, define

$$\dot{V}(\Phi) = \overline{\lim_{t \to 0}} \frac{1}{t} [V(u(t,\Phi)) - V(\Phi)].$$

V is said to be a Liapunov functional on a set G in $\mathscr{B}$ is V is continuous on $\overline{G}$ and if $\dot{V}(\Phi) \leq 0$ for every Φ in G. Furthermore, let $S = \{\Phi \text{ in } \overline{G} | \dot{V}(\Phi) = 0\}$ and let M be the largest invariant set in S for the dynamical system u.

Then it is possible to prove [9]

Theorem 2.1. Suppose u is a dynamical system on $\mathscr{B}$. If V is a Liapunov functional on G and the orbit $O^+(\varphi)$ belongs to G then $u(t,\varphi) \to S$ as $t \to \infty$. Furthermore, if $O^+(\varphi)$ belongs to a compact set of $\mathscr{B}$ then $u(t,\varphi) \to M$, and M is nonempty, compact and invariant.

This is one of the most general stability theorems available. Note that first of all, we always require the orbit to remain in G; secondly, that compactness of the orbit allows much more to be said about the set of points approached if S contains more than one element.

In the next examples, we attempt to illustrate the application of this general theorem. Note that the elements needed are:

(i) a dynamical system

(ii) a set $G \subset \mathcal{B}$

(iii) a Liapunov functional on G and, finally, perhaps

(iv) compactness of the orbits

3. A Problem of Nonexistence of Oscillations

Consider the network shown in Figure 1. In this circuit the section between 0 and 1 is a lossless transmission line with specific capacitance C_s and specific inductance L_s. The current i and the voltage v of this line are functions of ξ and t and satisfy the equations

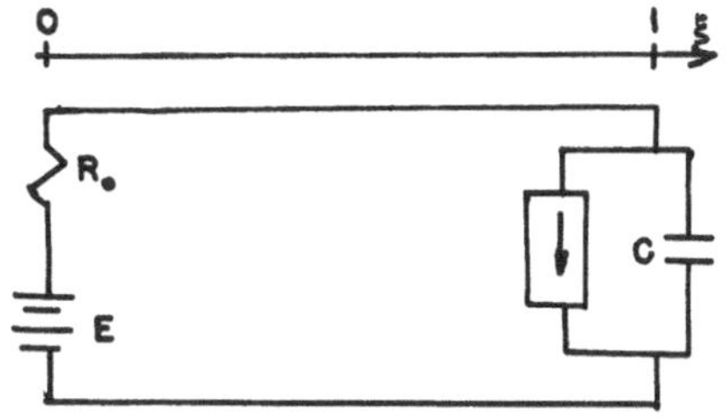

Figure 1

$$L_s \frac{\partial i}{\partial t} = -\frac{\partial v}{\partial \xi}, \qquad -C_s \frac{\partial v}{\partial t} = \frac{\partial i}{\partial \xi} \qquad 0 < \xi < 1, \quad t > 0. \tag{3.1}$$

The circuits at the ends of the line give rise to the boundary conditions

$$E = v_0 + R_0 i_0, \qquad C \frac{dv_1}{dt} + f(v_1) = i_1, \quad t > 0, \tag{3.2}$$

where $v_0(t) = v(0,t)$, $v_1(t) = v(1,t)$, $i_0(t) = i(0,t)$ and $i_1(t) = i(1,t)$. The function f which renders the problem nonlinear is pictured in Figure 2 and represents the general characteristic on an Esaki diode.

There has been considerable recent interest in circuits of this type, generally called flip-flops, particularly regarding the existence and nonexistence of oscillations. Moser [16], Brayton [2] and Brayton and Miranker [3] have considered increasingly sophisticated mathematical models for the study of such circuits, from lumped models to the present one. The equilibrium states of (3.1), (3.2) are given by

$$\begin{aligned} E &= v_1 + R_0 i_1, \\ i_1 &= f(v_1), \end{aligned} \tag{3.3}$$

and, as illustrated in Figure 2, we shall consider only the case of a unique equilibrium point, say (v^*,i^*). Translating the equilibrium state to the origin and denoting the new variables by the same notation yields

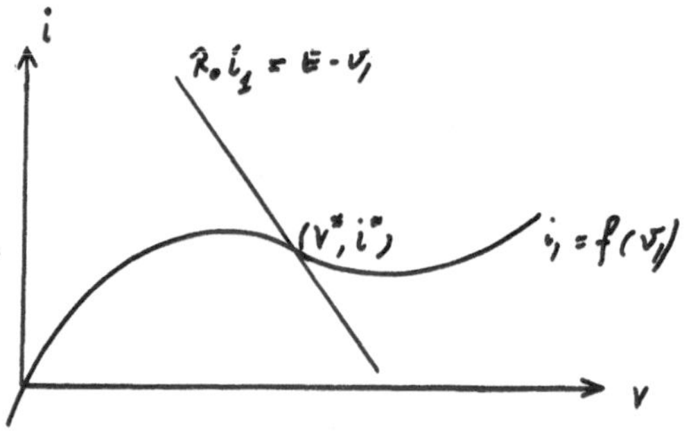

Figure 2

$$\begin{aligned} L_s \frac{\partial i}{\partial t} &= - \frac{\partial v}{\partial \xi}, \qquad 0 = v_0 + R_0 i_0, \\ -C_s \frac{\partial v}{\partial t} &= \frac{\partial i}{\partial \xi}, \qquad C \frac{dv_1}{dt} + g(v_1) = i_1, \end{aligned} \tag{3.4}$$

with $g(v_1) = f(v_1+v^*) - f(v^*)$, which is assumed continuously differentiable and globally lipschitzian.

The behavior of the solutions of (3.4) is far from obvious. What is desired is to determine conditions on the parameters that guarantee the global asymptotic stability of the solution; because of the nature of the circuit, the lossless transmission line, it is suspected that periodic oscillations are possible.

To study this problem with some mathematical care it is necessary to have an existence theorem which suggests the appropriate space in which the problem should be viewed; for this purpose it is fairly simple to prove [17]:

Theorem 3.1. For the system (3.7), let the initial conditions $i(\xi,0) = \hat{i}(\xi)$ and $v(\xi,0) = \hat{v}(\xi)$ belong to $C^1[0,1]$ and satisfy the consistency conditions

(i) $0 = -\hat{v}(0) - R_0\hat{i}(0)$

(ii) $0 = L_s\hat{i}'(0) + R_0 C_s \hat{v}'(0)$,

(iii) $\frac{C}{C_s}\hat{i}'(1) = -\hat{i}(1) + f(\hat{v}(1))$,

then there exists a unique solution $v(\xi,t)$, $i(\xi,t)$ in $C^1[0,1] \times C^1[0,\infty)$. Furthermore, this solution has the representation

$$\begin{aligned} v(\xi,t) &= \frac{1}{2}[\phi(\xi-\sigma t) + \psi(\xi+\sigma t)], \\ i(\xi,t) &= \frac{1}{2z}[\phi(\xi-\sigma t) - \psi(\xi+\sigma t)], \end{aligned} \qquad (3.5)$$

with $\sigma = \frac{1}{(L_s C_s)^{1/2}}$, $z = (\frac{L_s}{C_s})^{1/2}$.

This theorem yields a representation for the solutions which is very suggestive; through the use of this representation it is possible to reduce this problem to a more tractable one. Indeed, introducing (3.5) into (3.4), the wave equation is automatically satisfied and the boundary conditions become

$$\begin{aligned} v_1(t) + zi_1(t) &= -\psi_1(t - \tfrac{2}{\sigma})(\tfrac{z-R_0}{z+R_0}), \\ v_1(t) - zi_1(t) &= \psi_1(t), \\ c\frac{dv_1}{dt} + g(v_1) &= i_1. \end{aligned} \qquad (3.6)$$

Eliminating i_1 and ψ_1 then yields the neutral functional differential equation

$$C\frac{d}{dt}[v_1(t) + kv_1(t-r)] = -\frac{v_1(t)}{z} + \frac{k}{z}v_1(t-r) - g(v_1(t)) - kg(v_1(t-r)), \qquad (3.7)$$

where $r = \frac{2}{\sigma}$ and $k = \frac{R_0 - z}{R_0 + z}$. It is also simple to see that the given initial data $\hat{i}(\xi)$, $\hat{v}(\xi)$ in $C^1[0,1]$ completely determines the initial data $v_1 \in C^1[-r,0]$ for (1.7). Furthermore, it is not difficult to see that since $|k| < 1$ if

$\lim_{t \to \infty} v_1(t) = 0$, then $\lim_{t \to \infty} i(\xi,t) = 0$ and $\lim_{t \to \infty} v(\xi,t) = 0$ uniformly in ξ and that therefore oscillations will not exist.

The problem has then been reduced to the determination of conditions for the global asymptotic stability (3.7), which is rewritten for convenience of later computations as

$$\frac{d}{dt}[Dv_{1t}] = -[\frac{1}{Cz} + \frac{g(v_1(t))}{Cv_1(t)}]v_1(t) + [\frac{k}{Cz} - \frac{k}{C}\frac{g(v_1(t-r))}{v_1(t-r)}]v_1(t-r), \tag{3.8}$$

where $D\varphi = \varphi(0) + k\varphi(-r)$, $x_t(\theta) = x(t+\theta)$ with $-r \leq \theta \leq 0$. Cruz and Hale [10] have developed existence, uniqueness and continuous dependence results for this type of neutral functional differential equation.

Indeed, it should be noted that this is a functional differential equation of the neutral type of the type described in Example 3. Within this context and considering the application of the first part of Theorem 2.1 leads to

<u>Theorem 3.2</u>. If the D operator is a stable one and V is a Liapunov functional on $G = G_\rho = \{\varphi \in C: V(\varphi) < \rho\}$. Then, if $\dot{V}(\varphi) \leq -\omega(|D\varphi|) \leq 0$ with $\omega(s) > 0$ for $S > 0$, with ω continuous, then every solution of (1.3) approaches zero as $t \to \infty$.

The result is precisely the one expected as a generalization of the usual theorems for ordinary differential equations. Now, through the use of this theorem it is not too difficult to obtain some stability results for our problem. Indeed, it is possible to prove [17].

<u>Theorem 3.3</u>. If g satisfies the sector criterion

$$\sup_\sigma \left(\frac{g(\sigma)}{\sigma}\right) \leq \left(\frac{1-|k|}{1+|k|}\right)\frac{1}{z} + \inf_\sigma \left(\frac{g(\sigma)}{\sigma}\right),$$

and

$$\inf_\sigma \left(\frac{g(\sigma)}{\sigma}\right) \geq -\frac{1}{z}\left(\frac{1-|k|}{1+|k|}\right),$$

then the equilibrium solution $v_1 = 0$ of Equation (3.8) is globally asymptotically uniformly stable.

The proof of this theorem is straightforward, although the detailed computations are involved. In essence, the Liapunov functional $V(\varphi) = \frac{1}{2}[D\varphi]^2 + \alpha \int_{-r}^{0} \varphi^2(\theta)d\theta$ is used and conditions for the existence of a nonnegative α such that $\dot{V}(t,\varphi) \leq -\beta[D\varphi]^2$, $\beta > 0$, are determined. These conditions yield the sector criterion quoted in the theorem.

From what has been said above, these sector criteria naturally also imply the nonexistence of oscillations in the original problem. It is of interest to note that these criteria are sharp in the following sense. If the problem is linear, that is, $g(\sigma) = -\gamma\sigma$, then it is a simple exercise to determine that the condition $-\gamma \geq -\frac{1}{z}\left(\frac{1-|k|}{1+|k|}\right)$ is a necessary and sufficient condition for the non-existence of oscillations. But in the linear case, this is precisely the condition given by Theorem 3.3, which implies that a type of Aizerman conjecture is valid for this problem.

4. A Bifurcation Problem

A number of applications, especially those arising from chemical reactor stability problems [1] give rise to a problem of the following nature. Consider the partial differential equation

$$u_t = u_{xx} + \lambda f(u), \quad \lambda \geq 0, \quad 0 \leq x \leq \pi, \quad t > 0 \tag{4.1}$$

which satisfies the boundary and initial conditions

$$\begin{aligned} u(0,t) &= u(\pi,t) = 0, & t &\geq 0, \\ u(x,0) &= \phi(x), & 0 &\leq x \leq \pi \end{aligned} \tag{4.2}$$

where f is a given function defined on the real line, $f(0) = 0$, $uf(u) > 0$ for

$u \neq 0$ and $f(u)u^{-1} \to 0$ as $|u| \to \infty$. Assume for simplicity that f is C^2 smooth, odd and $\operatorname{sgn} f''(u) = -\operatorname{sgn} u$. With the given hypotheses $u \equiv 0$ is an equilibrium solution of this problem. For $\lambda = 0$ it is well known that this solution of the heat equation is stable in any usual meaning of the word, and the qualitative behavior of the solutions of (4.1), (4.2) is clear. What is of interest here is to determine how this picture changes as λ is allowed to increase from zero value; if the equilibrium solution $u \equiv 0$ loses its property of stability, do there appear any new equilibrium solutions which inherit this property? This problem has been investigated by Matkowsky [15] using formal asymptotic methods under hypothesis differing somewhat from these given here. The viewpoint here is to interpret (4.1), (4.2) as a dynamical system in an appropriate Banach space and to apply Liapunov methods of the type developed in [9, 11, 14]. Again, the details are omitted for the sake of the brevity of exposition. This specific application is more fully described in [5].

The first task here is to show that (4.1) - (4.2) defines a dynamical system. As a first step in this direction, consider the Banach space X of functions $\phi\colon [0,\pi] \to R$ continuously differentiable on $[0,\pi]$ with $\phi(0) = \phi(\pi) = 0$ and with norm $\|\phi\|_1 = \sup\{|\phi'(x)|\colon 0 \leq x \leq \pi\}$. Define also the norms $\|\phi\|_0 = \sup\{|\phi(x)|\colon 0 \leq x \leq \pi\}$ and $\|\phi\|_{W_2^1} = (\int_0^\pi \phi'(x)^2 dx)^{1/2}$, and note that $\|\phi\|_0 \leq \sqrt{\pi}\,\|\phi\|_{W_2^1} \leq \pi\|\phi\|_1$. Let $B_0(r)$ be open balls centered at zero with radius r in the $\|\ \|_0$ norm. Then it is possible to prove [5]:

<u>Theorem 4.1.</u> For any $\phi \in X$ and $\lambda \in [0,\infty)$, Equations (4.1), (4.2) have unique solutions $u(x,t;\,\phi,\lambda)$ denoted by $u(\phi,\lambda)(t) \in X$ defined for $0 < t < s(\phi,\lambda) \leq \infty$. Furthermore, if $u(\phi,\lambda)(t) \in B_0(r)$ for some r then $s(\phi,\lambda) = \infty$, the map $(t,\phi) \to u(\phi,\lambda)(t)$ defined for all $\phi \in X$ is a dynamical system in X with $\|\ \|_1$ and furthermore, the positive orbit $0^+(\phi,\lambda)$ of $u(\phi,\lambda)(t)$ is relatively compact in this space.

Note that, except for the hypothesis that the orbits are bounded in the

$\| \|_0$ norm, the theorem states that we are dealing with a dynamical system; furthermore, that the dynamical system is self-compactifying. This last property is precisely the expected result, given the smoothing properties of the heat equation which, this theorem states, are not affected by the nonlinearity.

Let us now define for every $\lambda \in [0,\infty)$ the Liapunov functional $V_\lambda(\phi) = \int_0^\pi \{\frac{1}{2}\phi(x)^2 - \lambda \int_0^{\phi(x)} f(\xi)d\xi\}dx$ for $\phi \in X$. Note that V_λ is continuous on X relative to $\| \|_1$ and $\| \|_{W_2^1}$ and that, given the assumptions it is not too difficult to see that $V_\lambda(\phi) \to \infty$ as $\|\phi\|_0 \to \infty$. Furthermore, it is of interest to see that $\frac{d}{dt} V_\lambda(u(\phi,\lambda)(t)) = -\int_0^\pi u_t^2(x,t; \phi,\lambda)^2 dx \leq 0$, for $0 < t < s(\phi,\lambda)$. These observations lead to

Theorem 4.2. For any $\phi \in X$ and $\lambda \in [0,\infty)$ the map $t,\phi \to u(\phi,\lambda)(t)$ is a dynamical system in X normed by $\| \|_1$. Furthermore, the positive orbit $0^+(\phi,\lambda)$ is relatively compact in this space.

Note that the use of the Liapunov functional was essential in proving global existence. But now, since the Liapunov function has already been constructed it is possible to conclude much more.

Indeed, all of the conditions for the entire Theorem 2.1 are satisfied. Note that the largest invariant set M within our context is the set of equilibrium solutions. Hence

Theorem 4.3. Every solution of (4.1) - (4.2) approaches an equilibrium solution in the norm $\| \|_1$.

Actually, much more can be said about the qualitative picture by analyzing the equilibrium solutions, which are the solutions of the two point boundary value problem

$$u''(x) + \lambda f(u(x)) = 0, \quad u(0) = u(\pi) = 0, \quad 0 \leq \lambda < \infty. \tag{4.3}$$

Using methods inspired by the work of Urabe [20] it is possible to prove

<u>Theorem 4.4.</u> Let $\lambda_n = \frac{n^2}{f'(0)}$, $n = 1,2,\ldots$. Then, for any $\lambda \in [\lambda_n,\infty)$ Equation (4.3) has two solutions $u_n^{\pm}(\lambda) \in B_0(r_0)$ with the properties that

(i) $u_n^{\pm}(\lambda_n) = 0$

(ii) $u_n^{\pm}(\lambda)$ have exactly $n + 1$ zeros in $[0,\pi]$

(iii) $u_n^{\pm}(\lambda)$ varies continuously in λ relative to the norm $\| \|_1$ with $\|u_n^{\pm}(\lambda)\|_1 \to \infty$ as $\lambda \to \infty$.

Furthermore, for any $\lambda \in [0,\infty)$, (4.1) have no equilibrium points in X other than the origin $u_0 \equiv 0$ and those elements $u_n^{\pm}(\lambda)$, $n \geq 1$, for which $\lambda_n < \lambda$.

It is then quite clear that if $\lambda \leq \lambda_1$ then for every $\phi \in X$ the corresponding solution $u(\phi,\lambda)(t) \to 0$ as $t \to \infty$, the convergence naturally being in the norm $\| \|_1$. The question arises, given $\phi \in X$ and $\lambda \in (\lambda_1,\infty)$ to which equilibrium point $u(\phi,\lambda)(t)$ will converge. Again, it is possible to answer, at least partially, this query by an appropriate analysis of the Liapunov functional. Indeed, we have

<u>Theorem 4.5.</u> For each integer $n \geq 1$, let $u_n^{\pm}(\lambda)$, $\lambda_n \leq \lambda < \infty$ be as in Theorem 4.4. Then for any $\lambda \in (\lambda_1,\infty)$ the origin $u_0 = 0$ is unstable. For any $\lambda \in [\lambda_1,\infty)$, $u_1^{\pm}(\lambda)$ is asymptotically stable and for $\lambda \in [\lambda_n,\infty)$, $n \geq 2$, $u_n^{\pm}(\lambda)$ is unstable. (All these assertions are valid in X normed by $\| \|_1$).

These five theorems give a rather clear picture of the qualitative behavior of the solutions. All solutions will, in general, approach either u_0 or $u_1^{\pm}(\lambda)$

5. The General Problem of Thermoelasticity

In the previous problem it was possible to find a Banach space in which the dynamical system was self-compactifying. It was this property that was heavily exploited and which is essential in the application of invariance principles. It is to be suspected that such self-compactifying properties can be expected in dynamical systems which arise from functional differential equations of the retarded type and partial differential equations of parabolic nature. For hyperbolic partial differential equations clearly this property would be very surprising. The example presented now is of hyperbolic nature, yet it is possible, through a little more work, to still apply the principle.

Elastic stability is usually discussed from strictly mechanical considerations; here the concern is with thermodynamic properties of elastic materials. More specifically, one may ask what effects the second law of thermodynamics has on the asymptotic stability of equilibrium of otherwise non-dissipative materials [7].

A material point is identified by $x = (x_1,x_2,x_3)$ in its state of equilibrium (no stresses, constant temperature $= \gamma_0$). The displacement field at some time t following an initial disturbance at time $t = 0$ is given by $u(x,t)$ and the temperature deviation by $T(x,t)$; $\rho(x)$ denotes the density at x in the equilibrium state. Let Ω be an open, bounded, connected set in E^3 which is properly regular [8]; let $\partial\Omega$ denote the boundary of Ω. The constitutive equations of thermoelasticity are taken then in the form

$$\rho\ddot{u}_i = (C_{ijk\ell})_{,j} - (m_{ij}T)_{,j}, \tag{5.1}$$

$$\rho C_0\dot{T} + m_{ij}\gamma_0\dot{u}_{i,j} = (K_{ij}T_{,j})_{,i}; \tag{5.2}$$

where body forces and heat sources have been excluded. In these equations $C_{ijk\ell} = C_{jik\ell} = C_{k\ell ij}$, $m_{ij} = m_{ji}$, $K_{ij} = K_{ji}$ and C_D, ρ, $C_{ijk\ell}$, m_{ij} and K_{ij} are assumed to be smooth functions of x.

Let now $t_0 > 0$. By a classical solution of the mixed initial-boundary value problem in $\Omega \times (0,t_0)$ we mean a pair (u,T) satisfying equation (5.1) and (5.2) together with the boundary conditions

$$u = 0 \quad \text{on} \quad \partial\Omega \times (0,t_0) \quad \text{(clamped boundary)}, \tag{5.3}$$

$$T = 0 \quad \text{on} \quad \partial\Omega \times (0,t_0) \quad \text{(constant temperature)}; \tag{5.4}$$

and with initial conditions

$$(u(x,0),\dot{u}(x,0),T(x,0)) = (u_0(x),\dot{u}_0(x),T_0(x)), \tag{5.5}$$

where $u_0(x)$, $\dot{u}_0(x)$ and $T_0(x)$ are given functions on Ω.

The generalized solutions of the mixed initial boundary value problem described above can be viewed on an appropriate Banach space as a dynamical system. Once this is done, the application of Theorem 2.1 permits us to draw immediate conclusions on the asymptotic behavior of the solutions of our problem.

Consider the Sobolev spaces $W_2^{(k)}(\Omega)$ and $W_{20}^{(k)}(\Omega)$, $k = 1,2,\dots$. Assume that

$$\text{ess inf } \rho(x) > 0, \text{ ess inf } C_D(x) > 0, \tag{5.6}$$

$$K_{ij}\xi_i\xi_j \geq C_1\xi_i\xi_i, \quad C_1 > 0 \quad \text{constant}, \tag{5.7}$$

(the second law of thermodynamics requires K_{ij} positive semidefinite at $x \in \Omega$; we make the stronger assumption of positive definiteness). Also for all $v_i \in W_{20}^{(1)}(\Omega)$

$$\int_\Omega C_{ijk\ell}v_{i,j}v_{k,\ell}dx \geq C_2 \int_\Omega v_{i,j}v_{i,j}dx, \quad C_2 > 0 \quad \text{constant} \tag{5.8}$$

Define now the spaces $H_0(\Omega) \approx W_{20}^{(1)}(\Omega) \times L_2(\Omega) \times L_2(\Omega)$ with norm $|(v_i,w_i,R)|_0^2 = \int_\Omega [\rho w_i w_i + C_{ijk\ell}v_{i,j}v_{k,\ell} + \frac{\rho C_D}{\gamma_0} R^2]dx$ and $H(\Omega) = W_{20}^{(1)}(\Omega) \times W_{20}^{(1)}(\Omega) \times W_{20}^{(1)}(\Omega)$. Define the map $P: H_0(\Omega) \overset{\text{onto}}{\rightarrow} H_1(\Omega)$ sending $(v_i,w_i,R) \in H_0(\Omega)$ onto $(u_i,v_i,T) \in H(\Omega) \subset H(\Omega)$ where $(u_i,T) \in W_{20}^{(1)}(\Omega) \times W_{20}^{(1)}(\Omega)$ is defined by the solution of the system

$$\int_\Omega C_{ijk\ell}u_{k,\ell}\theta_{i,j}dx = -\int_\Omega [\rho w_i\theta_i - m_{ij}T\theta_{i,j}]dx$$

$$\int_\Omega K_{ij}T_{,j}D_{,i}dx = -\int_\Omega [\rho C_D R + m_{ij}\gamma_0 v_{i,j}]D\,dx$$

for every $D_{,}\theta_i \in W_{20}^{(1)}(\Omega)$. The mapping P_m is linear, well-defined on $H_0(\Omega)$ and one to one. Hence, defining $P_m = P\circ P\circ\ldots\circ P$ let $H_m(\Omega)$ denote the range of the map P_m. It is clear that P_m^{-1} exists and maps $H_m(\Omega)$ onto $H_0(\Omega)$. Let $\psi \in H_m(\Omega)$ and define $|\psi|_m = |P_m^{-1}\psi|_0$. Then [6],

<u>Lemma 5.1.</u> H_m is a Banach space with norm $|\cdot|_m$. $H_0(\Omega) \supset H(\Omega) \supset \ldots \supset H_m(\Omega)$ algebraically and topologically. Furthermore, $H_m(\Omega)$ is dense in $H_\ell(\Omega)$ for $m > \ell$ and the imbedding $I: H_m(\Omega) \to H_\ell(\Omega)$ is compact.

Let us now define appropriately a <u>generalized solution</u> of our problem:

<u>Definition 5.1.</u> $(u_{i,}\dot{u}_{i,}T)$ will be called a <u>generalized solution</u> of (5.1) - (5.5) on $\Omega \times (0,t_0)$ if for all smooth test functions $(v_{i,}R)$ with compact support on Ω and v_i vanishing on $\Omega \times 0$

$$\begin{aligned}\int_0^{t_0}\int_\Omega \{(t-t_0)[\rho\dot{u}_i\ddot{v}_i - C_{ijk\ell}u_{k,\ell}\dot{v}_{i,j} + m_{ij}T\dot{v}_{i,j} + \\ + \frac{\rho C_D}{\gamma_0}T\dot{R} + m_{ij}u_{i,j}\dot{R}] + \rho\dot{u}_i\dot{v}_i + \rho\frac{C_D}{\gamma_0}TR + \\ + m_{ij}u_{i,j}R - \frac{1}{\gamma_0}\int_0^t (K_{ij}R_{,i})_{,j}T\,dt\}dxdt \\ = -t_0\int_\Omega [\rho\dot{u}_{0_i}v_i|_{t=0} + \frac{\rho C_D}{\gamma_0}T_0R|_{t=0} + m_{ij}u_{0_{i,j}}R|_{t=0}]dx.\end{aligned} \tag{5.9}$$

With this definition it follows that [6]:

<u>Theorem 5.1.</u> Under assumptions (5.1) - (5.3) the triple $(u_{i,}\dot{u}_{i,}T)$ describes a dynamical system on $H_m(\Omega)$, $m = 0,1,2,\ldots$, where $(u_{i,}\dot{u}_{i,}T)$ is the generalized solution to the equations of linear thermoelasticity satisfying equation (5.9). Furthermore, for t in $(0,t_0)$

$$|(u_i,\dot{u}_i,T)(t)|_m^2 + \frac{1}{\gamma_0}\int_0^t\int_\Omega K_{ij}T^{(m)}_{,i}T^{(m)}_{,j}\,dx\,d\tau = |(u_{i_0},\dot{u}_{i_0},\dot{T}_0)|_m^2 \tag{5.10}$$

where $T^{(m)}(x,t)$ denotes the generalized m^{th} derivative in time of $T(\dot{x},t)$.

The problem of termoelastic stability has now been put in a setting appropriate for the application of Theorem 2.1 which allows us to obtain stability results in a simple and direct manner.

For the trajectory $(u_i,\dot{u}_i,T)$ in $H_m(\Omega)$ define $P^\circ(u_i,\dot{u}_i,T) \equiv (\bar{u}_i,\dot{\bar{u}}_i,\bar{T})$. It follows from the definition of the map P that $(\bar{u}_i,\dot{\bar{u}}_i,\bar{T})$ is a dynamical system on $H_{m+1}(\Omega)$ with initial data $P^\circ(u_{0_i},\dot{u}_{0_i},T_0)$ in $H_{m+1}(\Omega)$ satisfying (5.9) and Theorem 5.1. Therefore, Theorem 2.1 and (5.10) imply that for any initial data $(u_{0_i},\dot{u}_{0_i},T_0)$ in $H_m(\Omega)$ the trajectory $(\bar{u}_i,\dot{\bar{u}}_i,\bar{T})(t)$ will lie in a bounded set of $H_m(\Omega)$ for all $t \geq 0$. Hence by Lemma 5.1 the trajectory $(\bar{u}_i,\dot{\bar{u}}_i,\bar{T})$ will lie in a compact set G of $H_\ell(\Omega)$, $\ell \leq m$. But then all the hypotheses of Theorem 2.1 are met with $\mathscr{B} = H_\ell(\Omega)$. For simplicity let $\ell = 1$ and $V = |(\bar{u}_i,\dot{\bar{u}}_i,\bar{T})|_1^2$. From (5.7) and (5.10) it immediately follows that $\dot{V} = \frac{-1}{\gamma_0}\int_\Omega K_{ij}\overset{(1)}{T}_{,i}\overset{(1)}{T}_{,j}\,dx \leq -c_3|(0,0,\bar{T})|_1^2$. The set S is then $S = \{(\bar{u}_i,\dot{\bar{u}}_i,\bar{T}) \in H_1(\Omega) | \bar{T} = 0\}$. The determination of M, the largest invariant set in S, which is not trivial, then leads to [18]:

<u>Theorem 5.2</u>. For any initial data $(u_{0_i},\dot{u}_{0_i},T_0)$ in $H_m(\Omega)$, $m \geq 1$, and under assumptions (5.6) - (5.7), $(u_i,\dot{u}_i,T)(t)$ approaches the set $M = \{(w_i,\dot{w}_i,Y)$ in $H_0(\Omega) | m_{ij}w_{i,j} = 0,\ Y = 0,\ \int_0^{t_0}\int_\Omega \{(t-t_0)[\rho\dot{w}_i\ddot{v}_i - C_{ijk\ell}w_{k,\ell}\dot{v}_{i,j}] + \rho\dot{w}_i\dot{v}_i\}dxdt =$ $-t_0\int_\Omega \rho\dot{w}_{0_i}\dot{v}_i|_{t=0}dx$ for all v_i test functions with compact support on Ω and vanishing on $\Omega\times 0$ in the norm of the space $H_0(\Omega)$ as $t \to \infty$.

It is of interest to note that in this case there is an infinity of solutions in the set M and that the use of the Liapunov functional allows a very nice characterization of them; they are the isothermal oscillations of the body, representing pure shear stresses. It should be noted that to obtain the needed compactification it is necessary for the problem to represent a dynamical system in two Banach spaces, here, for example, H_1 and H_0 with the imbedding of H_1 into H_0 completely continuous. The boundedness of the trajectories in H_1 then imply that the trajectory is in a compact set in H_0 and allows the application of the theorem. In this problem, which is linear, the generation of the H_n spaces is quite natural, they are velocity spaces. For nonlinear problems, unfortunately, this is far from easy.

6. Summary

In this brief lecture an attempt has been made to indicate the power and difficulties of application of Liapunov stability theory, with emphasis on the invariance principle. Looking back over the three examples, it is quite clear that the construction of the Liapunov functional is, in general, necessary to obtain the boundedness results required by a dynamical system. Once this functional is known, then if its derivative is negative definite in an appropriate domain, then only one equilibrium point will be stable. If the derivative is negative semidefinite, but the trajectory lies in a compact orbit, then the invariant subset of the set $\dot{V} = 0$ will be the set approached by the solutions. In the second example, the equations of motion were self-compactifying - in the last one they were not and one had to give initial conditions in a subspace which had the property that bounded set in it are compact in the larger space.

REFERENCES

[1] Admvuson, N. R. and L. R. Raymond; AICHE J., 11, 339-362, (1965).

[2] Brayton, R. K.; Quarterly Appl. Math., 24, (1966).

[3] Brayton, R. K. and W. L. Miranker; Arch. Rat. Mech. and Anal., 17, 61-73, (1964).

[4] Brockett, R. W.; IEEE Tr. Aut. Cont., 11, 596-606, (1966).

[5] Chafee, N. and E. F. Infante; Applicable Math., to appear.

[6] Dafermos, C. M.; Arch. Rat. Mech. and Anal., 29, 241-271, (1968).

[7] Eriksen, J. L.; Int. J. Solids and Structures, 2, 573-580, (1966).

[8] Fichera, G.; Lectures on Elliptic Boundary Differential Systems and Eigenvalue Problems, Springer-Verlag, 1965, p. 21.

[9] Hale, J. K.; J. Math. Anal. and Appl., 26, 39-59, (1969).

[10] Hale, J. K. and M. Cruz; J. Diff. Eqns., 7, 334-355, (1970).

[11] Hale, J. K. and E. F. Infante; Proc. Nat. Acad. Sci., 58, 405-409, (1967).

[12] Hale, J. K. and C. Imaz; Bul. Soc. Mat. Mex., 29-37, (1967).

[13] Holtzman, J. M.; Nonlinear System Theory, Prentice-Hall, (1970).

[14] LaSalle, J. P.; Int. Symp. Diff. Eqns. and Dym. Systems, Academic Press, 1967, p. 277.

[15] Matkowsky, B. J.; Bull. A. M. S., 76, 620-625, (1970).

[16] Moser, J.; Quarterly Appl. Math., 25, 1-9, (1967).

[17] Slemrod, M.; J. Math. Anal. and Appl., to appear.

[18] Slemrod, M. and E. F. Infante; Proc. IUTAM Symp. on Inst. Cont. Systems, Springer-Verlag, to appear.

[19] Sobolev, S. L.; Appl. of Funct. Anal. in Mat. Physics, Trans. Mat. Monographs, A. M. S., (1969).

[20] Urabe, M.; Army Math. Res. Center T. S. R. #437, (1963).

STABILITY OF DISSIPATIVE SYSTEMS WITH APPLICATIONS TO FLUIDS AND MAGNETOFLUIDS

E.M. Barston

Courant Institute of Mathematical Sciences
New York University, New York, New York

Abstract

An energy principle is presented which gives necessary and sufficient conditions for exponential stability for a useful class of continuous linear dissipative systems. The maximal growth rate Ω of an unstable system is shown to be the least upper bound of a certain functional, providing a variational expression for Ω. Applications to the problems of the stability of a stratified viscous incompressible fluid in a gravitational field and the resistive, viscous, incompressible magnetohydrodynamic sheet pinch are dicussed.

I. Introduction

In attempting to determine the stability characteristics of a given (usually nonlinear) physical system, one is often led to consider the stability of a derived (approximate) linear system. Perhaps it is known that the stability or instability of the original problem can in fact be inferred from the results obtained for the linearized problem; even if this information is not available, the lack of a general systematic method for the construction of Lyapunov functions often leaves one with no alternative, and so one proceeds with a study of the stability of the linear system, at least as a preliminary step in the solution of the problem.

Unfortunately, the solution of the derived linear problem itself is often formidable, even for autonomous systems, when the dimension is sufficiently large. This is particularly true for continuous systems where the linearized equations contain partial differential operators with spatially varying coefficients. Perhaps the best one can hope for in such cases is the existence of an energy principle which gives necessary and sufficient conditions for (exponential) stability. The existence of such an energy principle for determining the linear stability of the equilibrium states of a conservative dynamical system is well-known, and has been the cornerstone of almost every investigation of the stability of non-trivial equilibria in perfectly conducting, invicid, magneto-hydrodynamics [5],[6]. In 1903, Kelvin and Tate[8] proposed an extension of the energy principle to a class of real, finite-dimensional, dissipative linear systems (Kelvin and Tate did not prove their assertion; a proof using Lyapunov methods can be found in Ref. [7]). In recent years, the energy principle has been extended to a general class of continuous linear dissipative systems, and in the process, a maximum principle for the maximal growth rate of an unstable system has been obtained [1],[3]. We shall briefly discuss these developments and some applications in this paper. For a more complete discussion and further applications references [1]-[4] should be consulted.

We shall begin with a discussion of the problem of the gravitational stability of a stratified viscous incompressible fluid, which will serve to motivate as well as illustrate the theory. After developing the energy and maximum principles, we briefly dicuss the application of these results to the problem of the stability of the resistive, viscous, incompressible magneto-hydrodynamic sheet pinch.

II. Equations for a Viscous Incompressible Fluid in a Gravitational Field

Perhaps the most familiar example of a continuous dissipative system of the type we shall analyze is the problem of the gravitational stability of a stratified, viscous, incompressible fluid. Let us then consider such a fluid occupying a bounded region U (a simply connected open set) with surface ∂U, satisfying the following set of equations in U:

$$\nabla\cdot\vec{v} = 0 \tag{2.1}$$

$$\frac{\partial\rho}{\partial t} + \vec{v}\cdot\vec{\nabla}\rho = 0 \tag{2.2}$$

$$\rho\left\{\frac{\partial\vec{v}}{\partial t} + (\vec{v}\cdot\nabla)\vec{v}\right\} = -\vec{\nabla}p - \rho g\,\vec{e}_z + \nu\nabla^2\vec{v} \tag{2.3}$$

The quantity $\rho(\vec{x},t)$ denotes the mass density, $\vec{v}(\vec{x},t)$ the fluid velocity, $p(\vec{x},t)$ the scalar pressure, ν the viscosity (a positive constant), g the gravitational acceleration, and $\vec{e}_z$ the unit vector in the z-direction (assumed vertical). The equilibrium values of the fluid variables, denoted by the subscript o, are given as follows: $\vec{v}_o \equiv 0$; $\rho_o = \rho_o(z) > 0$ on $[z_1,z_2]$, $\rho_o \in C^1[z_1,z_2]$, where $z_1 \equiv \inf_{\vec{x}\in U} z$, $z_2 = \sup_{\vec{x}\in U} z$; and $p_o(z)$ is given by $p_o(z) = -g\int_{z_1}^{z}\rho_o(u)\,du + \text{const.}$ We linearize Eqs. (2.1)-(2.3) about the equilibrium state (in the sequel, the variables $\vec{v}$, p, and ρ without the subscript o will denote linearized quantities) and obtain, after introducing the (linear) displacement vector $\vec{\xi}(\vec{x},t) \equiv \int_0^t \vec{v}(\vec{x},\tau)\,d\tau + \vec{\xi}(\vec{x},0)$ where $\nabla\cdot\vec{\xi}(\vec{x},0) = 0$ and $\rho(\vec{x},0) = -\vec{\nabla}\rho_o\cdot\vec{\xi}(\vec{x},0)$,

$$\nabla\cdot\vec{\xi} = 0 \tag{2.4}$$

$$\rho_o\ddot{\vec{\xi}} - \nu\nabla^2\dot{\vec{\xi}} - g\frac{d\rho_o}{dz}\xi_z\,\vec{e}_z + \vec{\nabla}p = 0\ . \tag{2.5}$$

We take ∂U to be a rigid surface, so that the appropriate boundary condition is that $\vec{\xi}$ vanish on ∂U. We assume, of course, that all quantities are sufficiently smooth so that the indicated operations are well-defined; in particular, we consider the class of solutions $\vec{\xi}(\vec{x},t)$ of Eq. (2.5) such that $\vec{\xi}$ and $\dot{\vec{\xi}}$ are both in the class D and $\ddot{\vec{\xi}} \in C(\bar{\Omega})$ for each $t \geq 0$, where D is defined as the set of all functions $\vec{f}(\vec{x})$ with the properties that $\nabla\cdot\vec{f} = 0$ in U, $\vec{f} = 0$ on ∂U, and $\vec{f}$ is twice continuously differentiable on U. It is easy to see that the operators P,K, and H defined by $P\vec{\xi} \equiv \rho_o\vec{\xi}$, $K\vec{\xi} \equiv -\nu\nabla^2\vec{\xi}$, and $H\vec{\xi} \equiv -g\frac{d\rho_o}{dz}\xi_z\,\vec{e}_z$ are formally self adjoint on D with respect to the inner product $(\vec{f},\vec{g}) = \int_U \vec{f}^*\cdot\vec{g}\,d^3x$ ($\vec{f}^*$ denotes the complex conjugate of $\vec{f}$) and that P and K are positive. We note that $(\vec{\nabla}p,\vec{\xi}) = (\vec{\nabla}p,\dot{\vec{\xi}}) = 0$ for our solutions $\vec{\xi}$ of Eq. (2.5). This follows from the divergence theorem, since $\nabla\cdot\vec{\xi} = 0$ and $\vec{\xi}$ vanishes on ∂U.

III. The Energy and Maximum Principles

The preceeding problem is a special case of the more general system

$$P\ddot{\xi} + K\dot{\xi} + H\xi(t) + F_{\xi} = 0 , \qquad t \geq 0 \qquad (3.1)$$

where $\xi, \dot{\xi}, \ddot{\xi}$ and F_{ξ} are elements of an inner product space E for each fixed $t \geq 0$; P, K, and H are time-independent linear formally self-adjoint operators from E into E with domains of definition D_P, D_K, and D_H, respectively; $P \geq 0$ on D_P and $K \geq 0$ on D_K; and F_{ξ}, defined for each solution $\xi(t)$ of Eq. (3.1), has the property that $(F_{\xi}, \dot{\xi}) = (F_{\xi}, \xi) = 0$, $t \geq 0$. In the sequel, we restrict our attention to the class S of solutions $\xi(t)$ of Eq. (3.1) satisfying the following ten conditions:

$$\xi(t) \in D \equiv D_P \cap D_K \cap D_H , \qquad t \geq 0 \qquad (3.2)$$

$$\dot{\xi}(t) \in D_P \cap D_K , \qquad t \geq 0 \qquad (3.3)$$

$$\ddot{\xi}(t) \in D_P , \qquad t \geq 0 \qquad (3.4)$$

$$P\ddot{\xi} + K\dot{\xi} + H\xi + F_{\xi} = 0 , \qquad t \geq 0 \qquad (3.1)$$

$$\frac{d}{dt}(\dot{\xi}, P\dot{\xi}) = (\ddot{\xi}, P\dot{\xi}) + (\dot{\xi}, P\ddot{\xi}) , \qquad t \geq 0 \qquad (3.5)$$

$$\frac{d}{dt}(\dot{\xi}, P\xi) = (\ddot{\xi}, P\xi) + (\dot{\xi}, P\dot{\xi}) , \qquad t \geq 0 \qquad (3.6)$$

$$\frac{d}{dt}(\xi, P\xi) = (\dot{\xi}, P\xi) + (\xi, P\dot{\xi}) , \qquad t \geq 0 \qquad (3.7)$$

$$\frac{d}{dt}(\xi, K\xi) = (\dot{\xi}, K\xi) + (\xi, K\dot{\xi}) , \qquad t \geq 0 \qquad (3.8)$$

$$\frac{d}{dt}(\xi, H\xi) = (\dot{\xi}, H\xi) + (H\xi, \dot{\xi}) , \qquad t \geq 0 \qquad (3.9)$$

$$(F_{\xi}, \xi) = (F_{\xi}, \dot{\xi}) = 0 , \qquad t \geq 0 \qquad (3.10)$$

The class S may be thought of as the class of suitably "smooth" solutions of Eq. (3.1). Equations (3.5)-(3.9) are merely the usual rules for differentiating inner products; Eqs. (3.2)-(3.4) offer no restrictions on the solutions of Eq. (3.1) provided $D_P \supset D_K \supset D_H$, but become additional "smoothness" requirements should the above relation not hold.

The precise definition of the t-derivative $\dot{\xi}$ is not important in the sequel, provided that the usual rules for differentiating sums and products (of vectors and scalars) are valid. Thus one can think of $\dot{\xi}$ as being defined in the norm-topology of E, or if E is an n-fold Cartesian product of L_2-spaces (as is usually the case in applications), $\dot{\xi}$ can be taken to be the n-vector obtained by computing the partial derivative with respect to t of each of the n components of $\xi(t)$.

In addition to restricting the analysis to solutions $\xi(t) \in S$, we assume that H is bounded below on D and that $\inf_D \frac{(\eta,[\omega P+K]\eta)}{(\eta,\eta)} > 0$ for all $\omega > 0$. In the circumstance that $\inf_D \frac{(\eta,H\eta)}{(\eta,\eta)} < 0$, we define $\tilde{D} \equiv \{\eta | \eta \in D,\ (\eta,H\eta) < 0\}$, require $P > 0$ on $\tilde{D}$, set

$$Q_\eta \equiv \frac{1}{2}\left\{\left[\frac{(\eta,K\eta)^2}{(\eta,P\eta)^2} - 4\,\frac{(\eta,H\eta)}{(\eta,P\eta)}\right]^{1/2} - \frac{(\eta,K\eta)}{(\eta,P\eta)}\right\}$$

$$\text{for } \eta \in \tilde{D},\quad \Omega \equiv \sup_{\tilde{D}} Q_\eta,$$

$Y \equiv \{\phi |$ for each $\omega \in (0,\Omega)$, there exists $\xi(t) \in S$ and $\psi \in D_P \cap D_K$ such that $P\psi = 0$, $\xi(0) = \phi$, $\dot{\xi}(0) = \omega\phi+\psi\}$, and assume that $\sup_{Y\cap\tilde{D}} Q_\eta = \sup_{\tilde{D}} Q_\eta = \Omega$.

The stability of the solutions $\xi(t)$ of Eq. (3.1) will be discussed in terms of $\|\xi\| = (\xi,\xi)^{1/2}$. The function $\xi(t)$, defined for $t \geq 0$, is said to be exponentially stable if for every $\varepsilon > 0$, there exists a constant M_ε such that $\|\xi(t)\| \leq M_\varepsilon e^{\varepsilon t}$ for $t \geq 0$. If $\xi(t)$ is not exponentially stable, we say it is exponentially unstable. If every solution $\xi(t) \in S$ is exponentially stable, the system (3.1) is called exponentially stable.

With the preceeding definitions and hypothesis, we have the following theorem:

Theorem 1:

(A) Let $\inf_D \frac{(\eta,H\eta)}{(\eta,\eta)} > 0$. Then for each $\xi(t) \in S$, there exists a constant B such that $\|\xi(t)\| \leq B$ for all $t \geq 0$.

(B) Let $\inf_D \frac{(\eta,H\eta)}{(\eta,\eta)} = 0$. Then system (3.1) is exponentially stable.

(C) Let $\inf_D \frac{(\eta,H\eta)}{(\eta,\eta)} < 0$. Then the system is exponentially unstable with maximal growth rate Ω, i.e., given any $\omega \in (0,\Omega)$, there exists $\xi(t) \in S$ and a positive constant M such that $\|\xi(t)\| \geq M e^{\omega t}$ for all

$t \geq 0$, and given any $\xi(t) \in S$ and $\varepsilon > 0$, there exists a constant M_ε such that $\|\xi(t)\| \leq M_\varepsilon \, e^{(\Omega+\varepsilon)t}$, $t \geq 0$.

Proof: Let $\xi(t) \in S$. Then

$$\frac{d}{dt}\{(\dot{\xi},P\dot{\xi}) + (\xi,H\xi)\} = (P\ddot{\xi}+H\xi,\dot{\xi}) + (\dot{\xi},P\ddot{\xi}+H\xi)$$
$$= -2(\dot{\xi},K\dot{\xi}) - (F_\xi,\dot{\xi}) - (\dot{\xi},F_\xi)$$
$$= -2(\dot{\xi},K\dot{\xi}) \leq 0 , \qquad t \geq 0,$$

so that

$$(\dot{\xi},P\dot{\xi}) + (\xi,H\xi) \leq (\dot{\xi}_0,P\dot{\xi}_0) + (\xi_0,H\xi_0) , \quad t \geq 0 \quad (3.11)$$

where $\xi_o \equiv \xi(0)$, $\dot{\xi}_o \equiv \dot{\xi}(0)$. Let $\Delta \equiv \inf_D \frac{(\eta,H\eta)}{(\eta,\eta)}$. If $\Delta > 0$, Eq. (3.11) gives

$$\Delta\|\xi\|^2 \leq (\xi,H\xi) \leq (\dot{\xi}_0,P\dot{\xi}_0) + (\xi_0,H\xi_0), \quad t \geq 0 \quad (3.12)$$

which proves (A). Let $\omega > 0$, $\xi(t) \in S$, and set $\zeta(t) \equiv \xi(t)e^{-\omega t}$, $t \geq 0$. Then $\xi(t) = \zeta(t)e^{\omega t}$, and a straightforward calculation yields

$$P\ddot{\zeta} + K_\omega\dot{\zeta} + H_\omega\zeta + f_\zeta = 0 , \qquad t \geq 0 \quad (3.13)$$

where $K_\omega \equiv 2\omega P+K$, $H_\omega \equiv \omega^2P+ \omega K+H$, and $f_\zeta \equiv F_\xi \, e^{-\omega t}$, so that $(f_\zeta,\zeta) = (f_\zeta,\dot{\zeta}) = 0$ for $t \geq 0$. By analogy with the derivation of Eq. (3.11) we have

$$(\dot{\zeta},P\dot{\zeta}) + (\zeta,H_\omega\zeta) \leq (\dot{\zeta}_0,P\dot{\zeta}_0) + (\zeta_0,H_\omega\zeta_0) , \quad t \geq 0 \quad (3.14)$$

Let $\Delta = \inf_D \frac{(\eta,H\eta)}{(\eta,\eta)} = 0$. Then since $\inf_D \frac{(\eta,[\omega P+K]\eta)}{(\eta,\eta)} > 0$, we conclude that $\Delta_\omega \equiv \inf_D \frac{(\eta,H_\omega\eta)}{(\eta,\eta)} > 0$, so that Eq. (3.14) implies

$$\|\xi(t)\| = \|\zeta(t)\|e^{\omega t} \leq \left[\frac{(\dot{\zeta}_0,P\dot{\zeta}_0)+(\zeta_0,H\zeta_0)}{\Delta_\omega}\right]^{1/2} e^{\omega t}, \; t \geq 0$$

which holds for any $\omega > 0$. Thus statement (B) is verified. Now suppose that $\Delta < 0$. Then $\tilde{D}$ is nonempty, and for each $\eta \in \tilde{D}$, $Q_\eta > 0$, so that $\Omega > 0$. Let $\omega \in (0,\Omega)$. Since $\sup_{Y \cap \tilde{D}} Q_\eta = \Omega$, there exists $\phi \in Y$ such that

$$\omega < Q_\phi \leq \Omega, \text{ and a } \xi(t) \in S \text{ such that } \xi_0 = \phi ,$$

$$\dot{\xi}_0 = \omega\phi+\psi , \text{ where } P\psi = 0 . \text{ Set } \zeta(t) \equiv \xi(t)e^{-\omega t} .$$

Then $\zeta_0 = \phi$, $\dot{\zeta}_0 = \dot{\xi}_0-\omega\xi_0 = \psi$, and Eq. (3.14) yields

$$\Delta\|\zeta(t)\|^2 \leq (\phi,H_\omega\phi) \ , \qquad t \geq 0 \quad (3.15)$$

The quadratic function $g(\alpha) \equiv (\phi,H_\alpha\phi)$ is a strictly increasing function of α for $0 < \alpha < \infty$ and vanishes for $\alpha = Q_\phi$; thus $(\phi,H_\omega\phi) < 0$. We therefore conclude from Eq. (3.15) that

$$\|\xi\| = \|\zeta\| e^{\omega t} \geq \left[\frac{(\phi,H_\omega\phi)}{\Delta}\right]^{1/2} e^{\omega t} \ , \qquad t \geq 0 \ .$$

Thus the growth rate Ω can be approached arbitraily closely for some $\xi(t) \in S$. Finally, suppose that Ω is finite and let $\varepsilon > 0$. Since $\inf_D \frac{(\eta,[\omega P+K]\eta)}{(\eta,\eta)} > 0$ for $\omega > 0$, it follows that $\Delta_{\Omega+\varepsilon} = \inf_D \frac{(\eta,H_{\Omega+\varepsilon}\eta)}{(\eta,\eta)} > 0$. Let $\xi(t) \in S$ and set $\zeta(t) = \xi(t)e^{-(\Omega+\varepsilon)t}$. Then Eq. (3.14) gives

$$\|\xi\| = \|\zeta\| e^{(\Omega+\varepsilon)t} \leq \left[\frac{(\dot{\zeta}_o,P\dot{\zeta}_o)+(\zeta_o,H_{\Omega+\varepsilon}\zeta_o)}{\Delta_{\Omega+\varepsilon}}\right]^{1/2} e^{(\Omega+\varepsilon)t} , \quad t \geq 0 \ ,$$

which completes the proof.

The derivation of the energy principle given herein has the advantage of being free from any assumptions of completeness imposed on the eigenfunctions of the linear system; in fact, the results are valid for systems with no proper eigenfunctions. This is important in applications to systems with a continuous spectrum. We have basically made the much weaker assumption that the system (3.1) admits smooth solutions for smooth initial data, and do not require the existence of any solutions of the form $\xi(t) = \eta e^{\omega t}$, where η is independent of t. It should be clear that in general, Ω will not lie in the discrete spectrum, i.e., the theorem only guarantees that the growth rate Ω can be approached arbitrarily closely, but does not imply that it can actually be achieved.

IV. Applications

The energy and maximum principles of Theorem 1 are applicable to any system satisfying an equation of the form (3.1) and the associated hypothesis imposed in Sec. III. (It should be observed from the proof of Theorem 1 that relatively little of that hypothesis is required to prove exponential stability once $H \geq 0$ on D_H is known; the entire hypothesis was used, however, in the proof of the instability results and the maximum principle). There are two approaches to the rigorous application of the energy and maximum principles to a given problem. The first, and usually most difficult, requires an existence theorem guaranteeing the existence of the required smooth solutions for smooth initial data. The second approach, applicable to unstable systems where the maximal growth rate Ω lies in the discrete spectrum, is to demonstrate the existence of an eigenvector η (independent of t) such that $\xi(t) = \eta e^{\Omega t}$ is a solution of (3.1). This approach is valid for the resistive sheet pinch [2]. It is to be expected, however, that in most applications the investigator will simply assume that his system is well-behaved and that the energy and maximum principles apply. If the system is based on sound physical principles and the equilibrium data are sufficiently smooth, one would generally expect smooth solutions for smooth initial data. Then the only problem remaining is the choice of the domain $D = D_P \cap D_K \cap D_H$. A guiding principle here is to take D to be the "maximal" linear manifold satisfying the conditions that P,K, and H are all well-defined and formally self-adjoint on D (P is formally self-adjoint on D if and only if $(\eta, P\zeta) = (P\eta, \zeta)$ for all $\eta, \zeta \in D$) and that $P\eta$, $K\eta$, and $H\eta$ are reasonably smooth for all $\eta \in D$. Of course we require $P \geq 0$ and $K \geq 0$ on D. Returning to the problem discussed in Sec. II, we identify P with ρ_o, K with $-\nu\nabla^2$, and H with $-g \frac{d\rho_o}{dz} (\vec{e}_z \cdot)\vec{e}_z$. Due to the side condition (2.4) and the boundary condition $\vec{\xi} \equiv 0$ on ∂U, we take D to be the linear manifold of all vector functions $\vec{f}(\vec{x})$ such that $\nabla \cdot \vec{f} \equiv 0$ in U, $\vec{f} \equiv 0$ on ∂U, $\vec{f}$ is twice continuously differentiable in U, and the functions defined by the first and second partials of $\vec{f}$ can be extended to ∂U so that they are continuous on the closure of U. For $\eta = \vec{f} \in D$, we have $(\eta, H\eta) = - g \int_U \frac{d\rho_o}{dz} |f_z|^2 d^3x$; thus if $\frac{d\rho_o}{dz} \leq 0$ on U, $H \geq 0$ on D and we have exponential stability. If, on the other hand, $\frac{d\rho_o}{dz} > 0$ on some open sphere in U, then we can choose an $\eta \in D$ such that $(\eta, H\eta) < 0$, and we then "conclude" that the system is exponentially unstable with the maximal growth rate $\Omega = \sup_{\tilde{D}} Q_\eta$. The maximal growth rate Ω will of course depend on the viscosity ν, the mass density ρ_o, and the domain U.

The remainder of this section will be devoted to a brief discussion of the application of the energy principle to the resistive viscous, incompressible, magnetohydrodynamic sheet pinch. A detailed discussion can be found in [2]. (For an application to the electrohydrodynamic Rayleigh-Taylor bulk instability [9], see [4]).

We consider an infinite horizontal layer of an incompressible, viscous fluid satisfying the usual incompressible magnetohydrodynamic equations with a viscosity term added to the equation of motion, except for a simple "Ohm's Law" of the form $\vec{E} + \vec{v} \times \vec{B} = \eta\vec{J}$ and the addition of a conservation equation $\frac{\partial\eta}{\partial t} + \nabla\cdot(\eta\vec{v}) = 0$ for the resistivity η. The equilibrium quantities are assumed to be functions of the vertical coordinate z only, with the equilibrium fluid velocity identically zero, and the equilibrium magnetic field $\vec{B}_o(z)$ horizontal. The boundaries of the fluid (located at z = 0 and z = a) are assumed to be rigid, perfect insulators. The system equations require that the equilibrium electric field $\vec{E}_o$ be constant and horizontal, while $\vec{B}_o(z)$ and $\eta_o(z)$ are related by

$$\vec{B}_o(z) = \vec{B}_o(0) + \eta_o\vec{E}_o \times \vec{e}_z \int_0^z \eta_o^{-1}(u)\,du \ ,$$

where $\vec{B}_o(0)$ is a constant horizontal magnetic field and μ_o is the permeability of free space (mks units). The system equations are linearized about the equilibrium and the linearized variables are Fourier analyzed in the horizontal plane. After a great leal of algebra, the following 2 × 2 matrix equation is obtained, which determines the stability of the system:

$$P\ddot{\xi} + K\dot{\xi} + H\xi = 0 \ , \qquad (4.1)$$

where $\xi = \begin{pmatrix} \xi_1(z,t) \\ \xi_2(z,t) \end{pmatrix}$ with ξ_1 the Fourier coefficient of the z component of the perturbed displacement vector and ξ_2 the Fourier coefficient of the z component of the perturbed magnetic field; the 2 × 2 matrix operators P, K, and H have the form

$$P = \begin{pmatrix} L_1 & 0 \\ 0 & 0 \end{pmatrix}, \quad K = \begin{pmatrix} L_2 & 0 \\ 0 & 0 \end{pmatrix} + B_1, \quad H = \begin{pmatrix} 0 & 0 \\ 0 & L_3 \end{pmatrix} + B_2$$

where L_1 and L_3 are second-order linear differential operators in z, L_2 is a fourth-order differential operator, and B_1 and B_2 are 2×2 Hermitian matrix operators whose elements are continuous functions of z on [0,a] (we assume that all equilibrium quantities are twice continuously differentiable functions of z on [0,a]). Consideration of the boundary conditions and smoothness requirements leads us to require that for each $t \geq 0$, $\xi_1(z,t) \in D_1 \equiv \{f(z) | f \in C^4[0,a],\ f(0) = f'(0) = f(a) = f'(a) = 0\}$ and $\xi_2(z,t) \in D_2 \equiv \{f(z) | f \in C^2[0,a],\ f'(a)+kf(a) = 0 = f'(0)-kf(0)\}$, where k denotes the magnitude of the horizontal wave number vector. Thus we take $D = D_1 \times D_2$, and find that P,K, and H are all formally self-adjoint on D with $P \geq 0$ and $K > 0$. The energy principle is applicable (here $F_\xi \equiv 0$), and the result is that unless η_o is a constant, the pinch ($\vec{E}_o \neq 0$) is always exponentially unstable (for sufficiently small k).

The theory of Sec. III leads us to expect that if the sheet pinch is unstable at the wave number k, then the maximal growth rate of perturbations with this wave number will be given by $\Omega(k) = \sup_{\tilde{D}} Q_\eta$. (The maximal growth rate for arbitrary disturbances, i.e., disturbances of arbitrary wave number, would then be given by $\sup \Omega(k)$, where the supremum is over all k for which $\Omega(k) > 0$.) We now show that the maximal growth rate $\Omega(k)$ is actually achieved for the unstable (at wave number k) sheet pinch, i.e., we demonstrate the existence of a nonzero eigenvector $\psi \in D$ such that $\psi e^{\Omega t}$ satisfies Eq. (4.1). Let $k > 0$, and suppose that $\inf_D \frac{(\xi,H\xi)}{(\xi,\xi)} < 0$ (i.e., the system is unstable for wave number k). The operators L_1 and L_2 are strictly positive on D_1, and $B_1 \geq 0$ on $E \equiv \mathcal{L}_2[0,a] \times \mathcal{L}_2[0,a]$, so that $F(\omega) \equiv \inf_D \frac{(\xi,H_\omega\xi)}{(\xi,\xi)}$, $0 \leq \omega < \infty$, is strictly increasing on $[0,\infty)$. Now $\Omega = \sup_{\tilde{D}} Q_\eta > 0$, and we have $F(\omega) < 0$ on $[0,\Omega)$, $F(\Omega) \geq 0$. The operator L_3 has a positive compact Hermitian inverse K defined on $\mathcal{L}_2[0,a]$ such that

$K(\mathcal{L}_2[0,a]) \subset C[0,a]$, $L_3K = I$ on $C[0,a]$, and $KL_3 = I$ on D_2. For each $\omega > 0$, the operator $\omega^2 L_1 + \omega L_2$ has a positive compact Hermitian inverse K_ω defined on $\mathcal{L}_2[0,a]$ such that $K_\omega(\mathcal{L}_2[0,a]) \subset C[0,a]$, $(\omega^2 L_1 + \omega L_2)K_\omega = I$ on $C[0,a]$, $K_\omega(\omega^2 L_1 + \omega L_2) = I$ on D_1, and K_ω is continuous in ω on $(0,\infty)$. Thus for $\omega > 0$, $S_\omega \equiv \begin{pmatrix} \omega^2 L_1 + \omega L_2 & 0 \\ 0 & L_3 \end{pmatrix}$ with domain D admits the positive compact Hermitian inverse $T_\omega \equiv \begin{pmatrix} K_\omega & 0 \\ 0 & K \end{pmatrix}$ from E into E such that T_ω is continuous in ω, $T_\omega S_\omega = I$ on D, $S_\omega T_\omega = I$ on $C[0,a] \times C[0,a]$, and $T_\omega(E) \subset C[0,a] \times C[0,a]$. For $\omega > 0$, let $r_\omega \equiv T_\omega^{1/2}$, $B_\omega \equiv -\omega B_1 - B_2$, and $G(\omega) \equiv \inf_E \frac{(\phi,[I - r_\omega B_\omega r_\omega]\phi)}{(\phi,\phi)}$. Note that for $\zeta \in D$, $(\phi,[I - r_\omega B_\omega r_\omega]\phi) = (\zeta, H_\omega \zeta)$, where $\phi \equiv r_\omega S_\omega \zeta$. Therefore $F(\omega) < 0$ on $[0,\Omega)$ implies $G(\omega) < 0$ on $[0,\Omega)$, and since $\overline{r_\Omega S_\Omega(D)} = E$, $F(\Omega) \geq 0$ implies $G(\Omega) \geq 0$. It therefore follows from the continuity of $G(\omega)$ on $(0,\infty)$ that $G(\Omega) = 0$, i.e., $\sup_E \frac{(\phi, r_\Omega B_\Omega r_\Omega \phi)}{(\phi,\phi)} = 1$. The operator $r_\Omega B_\Omega r_\Omega$ is compact and Hermitian, so that there exists $\zeta \in E$, $\|\zeta\| = 1$, such that $\zeta = r_\Omega B_\Omega r_\Omega \zeta$. Hence $\psi \equiv r_\Omega \zeta = T_\Omega B_\Omega \psi \in C[0,a] \times C[0,a]$, so that $B_\Omega \psi \in C[0,a] \times C[0,a]$, which implies that $\psi = T_\Omega B_\Omega \psi \in D$. Therefore $S_\Omega \psi = S_\Omega T_\Omega B_\Omega \psi = B_\Omega \psi$, i.e., $H_\Omega \psi = 0$, and the proof is complete.

Acknowledgment

The work presented here was supported by the Magneto-Fluid Dynamics Division, Courant Institute of Mathematical Sciences., New York University, under U.S. Air Force Grant AFOSR-71-2053.

Bibliography

1. Barston, E.M., Comm. Pure and Appl. Math. 22, 627 (1969).
2. Barston, E.M., Phys. Fluids 12, 2162 (1969).
3. Barston, E.M., J. Fluid Mech. 42, 97 (1970).
4. Barston, E.M., Phys. Fluids 13, 2876 (1970).
5. Bernstein, I.B., Frieman, E.A., Kruskal, M.D., and Kulsrud, R.M., Proc. Roy. Soc. A 244, 17 (1958).
6. Chandrasekhar, S., Hydrodynamic and Hydromagnetic Stability, Chap. 14 (Oxford University Press, 1961).

7. Chetaev, N.G., The Stability of Motion, Chap. 5 (Pergamon Press, London 1961).

8. Thompson, W., (Lord Kelvin), and Tait, P.G., Treatise on Natural Philosophy, Part I, Secs. 339-345 (Cambridge University Press, London, 1903).

9. Turnbull, R.J., and Melcher, J.R., Phys. Fluids 12, 1160 (1969).

Lecture Notes in Physics

Bisher erschienen / Already published

Vol. 1: J. C. Erdmann, Wärmeleitung in Kristallen, theoretische Grundlagen und fortgeschrittene experimentelle Methoden. 1969. DM 20,–

Vol. 2: K. Hepp, Théorie de la renormalisation. 1969. DM 18,–

Vol. 3: A. Martin, Scattering Theory: Unitarity, Analyticity and Crossing. 1969. DM 16,–

Vol. 4: G. Ludwig, Deutung des Begriffs physikalische Theorie und axiomatische Grundlegung der Hilbertraumstruktur der Quantenmechanik durch Hauptsätze des Messens. 1970. DM 28,–

Vol. 5: M. Schaaf, The Reduction of the Product of Two Irreducible Unitary Representations of the Proper Orthochronous Quantummechanical Poincaré Group. 1970. DM 16,–

Vol. 6: Group Representations in Mathematics and Physics. Edited by V. Bargmann. 1970. DM 24,–

Vol. 7: R. Balescu, J. L. Lebowitz, I. Prigogine, P. Résibois, Z. W. Salsburg, Lectures in Statistical Physics. 1971. DM 18,–

Vol. 8: Proceedings of the Second International Conference on Numerical Methods in Fluid Dynamics. Edited by M. Holt. 1971. DM 28,–

Vol. 9: D. W. Robinson, The Thermodynamic Pressure in Quantum Statistical Mechanics. 1971. DM 16,–

Vol. 10: J. M. Stewart, Non-Equilibrium Relativistic Kinetic Theory. 1971. DM 16,–

Vol. 11: O. Steinmann, Perturbation Expansions in Axiomatic Field Theory. 1971. DM 16,–

Vol. 12: Statistical Models and Turbulence. Edited by M. Rosenblatt and C. Van Atta. 1972. DM 28,–

Vol. 13: M. Ryan, Hamiltonian Cosmology. 1972. DM 18,–

Vol. 14: Methods of Local and Global Differential Geometry in General Relativity. Edited by D. Farnsworth, J. Fink, J. Porter and A. Thompson. 1972. DM 18,–

Vol. 15: M. Fierz, Vorlesungen zur Entwicklungsgeschichte der Mechanik. 1972. DM 16,–

Vol. 16: H.-O. Georgii, Phasenübergang 1. Art bei Gittergasmodellen. 1972. DM 18,–

Vol. 17: Strong Interaction Physics. Edited by W. Rühl and A. Vancura. 1973. DM 28,–

Vol. 18: Proceedings of the Third International Conference on Numerical Methods in Fluid Mechanics, Vol. I. Edited by H. Cabannes and R. Temam. 1973. DM 18,–

Vol. 19: Proceedings of the Third International Conference on Numerical Methods in Fluid Mechanics, Vol. II. Edited by H. Cabannes and R. Temam. 1973. DM 26,–

Vol. 20: Statistical Mechanics and Mathematical Problems. Edited by A. Lenard. 1973. DM 22,–

Vol. 21: Optimization and Stability Problems in Continuum Mechanics. Edited by P. K. C. Wang. 1973. DM 16,–

Beschaffenheit der Manuskripte

Die Manuskripte werden photomechanisch vervielfältigt; sie müssen daher in sauberer Schreibmaschinenschrift mit ausreichend großer Type geschrieben sein. Handschriftliche Formeln bitte nur mit schwarzer Tusche eintragen. Notwendige Korrekturen sind bei dem bereits geschriebenen Text entweder durch Überkleben des alten Textes vorzunehmen oder aber müssen die zu korrigierenden Stellen mit weißem Korrekturlack abgedeckt werden. Die reproduktionsfähigen Abbildungen (in Originalgröße) sollen in den Text eingeklebt werden. Falls das Manuskript oder Teile desselben neu geschrieben werden müssen, ist der Verlag bereit, dem Autor bei Erscheinen seines Bandes einen angemessenen Betrag zu zahlen. Die Autoren erhalten 50 Freiexemplare.

Zur Erreichung eines möglichst optimalen Reproduktionsergebnisses ist es erwünscht, daß bei der vorgesehenen Verkleinerung der Manuskripte der Text auf einer Seite in der Breite möglichst 18 cm und in der Höhe 26,5 cm nicht überschreitet. **Entsprechende Satzspiegelvordrucke werden vom Verlag gern auf Anforderung zur Verfügung gestellt.**

Manuskripte, in englischer, deutscher oder französischer Sprache abgefaßt, sind einzureichen bei: Springer-Verlag, 6900 Heidelberg, Postfach 1780.

Cette série a pour but de donner des informations rapides, de niveau élevé, sur des développements récents en physique, aussi bien dans la recherche que dans l'enseignement supérieur. On prévoit de publier.

1. des versions préliminaires de travaux originaux et de monographies
2. des cours spéciaux portant sur un domaine nouveau ou sur des aspects nouveaux de domaines classiques
3. des rapports de séminaires
4. des conférences faites lors de congrès ou de colloques

En outre il est prévu de publier dans cette série, si la demande le justifie, des rapports de séminaires et des cours multicopiés ailleurs mais déjà épuisés.

Dans l'intérêt d'une diffusion rapide, les contributions auront souvent un caractère provisoire; le cas échéant, les démonstrations ne seront données que dans les grandes lignes. Les travaux présentés pourront également paraître ailleurs. Une réserve suffisante d'exemplaires sera toujours disponible. En permettant aux personnes intéressées d'être informées plus rapidement, les éditeurs Springer espèrent, par cette série de «prépublications», rendre d'appréciables services aux instituts de physique. Les annonces dans les revues spécialisées, les inscriptions aux catalogues et les copyrights rendront plus facile aux bibliothèques la tâche de réunir une documentation complète.

Présentation des manuscrits

Les manuscrits, étant reproduits par procédé photomécanique, doivent être soigneusement dactylographiés type assez grand. Il est recommandé d'écrire à l'encre de Chine noire les formules non dactylographiées. Les corrections nécessaires doivent être effectuées soit par collage du nouveau texte sur l'ancien soit en recouvrant les endroits à corriger par du vernis correcteur blanc. Les illustrations; en dimension originale, préparées pour reproduction sont à insérer dans le texte. S'il s'avère nécessaire d'écrire de nouveau le manuscrit, soit complètement, soit en partie, la maison d'édition se déclare prête à verser à l'auteur, lors de la parution du volume, le montant des frais correspondants. Les auteurs recoivent 50 exemplaires gratuits.

Pour obtenir une reproduction optimale il est désirable que le texte dactylographié sur une page ne dépasse pas 26,5 cm en hauteur et 18 cm en largeur. **Sur demande la maison d'edition met à la disposition des auteurs du papier spécialement préparé.**

Les manuscrits en anglais, allemand ou français peuvent être adressés à Springer-Verlag, 6900 Heidelberg, Postfach 1780.

Springer-Verlag, D-1000 Berlin 33, Heidelberger Platz 3
Springer-Verlag, D-6900 Heidelberg 1, Neuenheimer Landstraße 28-30
Springer-Verlag, 175 Fifth Avenue, New York, NY 10010/USA